LOUIS RUDD

WITH SEAN RAYMENT

ENDURANCE

SAS SOLDIER. POLAR ADVENTURER.
DECORATED LEADER.

PAN BOOKS

First published 2020 by Macmillan

This paperback edition first published 2021 by Pan Books
an imprint of Pan Macmillan
The Smithson, 6 Briset Street, London EC1M 5NR
EU representative: Macmillan Publishers Ireland Ltd,
Mallard Lodge, Lansdowne Village, Dublin 4
Associated companies throughout the world
www.panmacmillan.com

ISBN 978-1-5290-3176-8

Copyright © Louis Rudd 2020

The right of Louis Rudd to be identified as the
author of this work has been asserted by him in accordance
with the Copyright, Designs and Patents Act 1988.

All photographs are from the author's personal collection, except for p. 2 (bottom left)
© Henry Worsley, p. 3 (bottom left) © Mark Wood, and p. 12 (top) © René Koster

1 3 5 7 9 8 6 4 2

A CIP catalogue record for this book is available from the British Library.

Printed and bound by CPI Group (UK) Ltd, Croydon, CR0 4YY

Visit **www.panmacmillan.com** to read more about all our books
and to buy them. You will also find features, author interviews and
news of any author events, and you can sign up for e-newsletters
so that you're always first to hear about our new releases.

ENDURANCE

Captain Louis Rudd was a Royal Marine Commando for six years before joining the SAS, in which he served for over two decades. He is a veteran of military campaigns in Northern Ireland, Kosovo, Afghanistan and Iraq and has also taken part in four major polar expeditions. His first, the Scott Amundsen Centenary Race Expedition, involved two teams of soldiers who followed the same routes taken a hundred years earlier by the two greatest explorers of the era. He then led the SPEAR17 Expedition – an Army Reserve team crossing of Antarctica – for which he was awarded an MBE. He led another team traversing Greenland in 2018, and later that year he made history as the first Briton ever to complete a solo, unsupported crossing of Antarctica. *Endurance* is his first book.

CONTENTS

CONTENTS

To my best friend, mother of my three wonderful children and wife, Lucy. No man could ask for a more loyal, loving and supportive partner. Without her strength and blessing none of this would have been possible.

To Henry Worsley, for inspiring me, mentoring me in the dark arts of polar travel, and providing me with a true appreciation of polar history and a deep love of Antarctica. I owe him everything. Onwards.

PROLOGUE

If you're going through hell, keep going.

WINSTON CHURCHILL

It was 29 November 2018 and for four long days I had been trapped in a whiteout. I was skiing blind, but in the complete opposite of total darkness. It was light but I could see nothing beyond the length of my arm, not the sky above nor the ice beneath my feet. There was no wind and no sound apart from the endless crunching of ice under my skis. It was the twenty-seventh day of my attempt to complete a 920-mile, solo unsupported traverse of the Antarctic landmass – something no one had managed to achieve before without wind assistance from a kite or food resupplies. Antarctica had defeated more experienced men than me.

I had already carved my way through more than 500 miles of polar ice when disaster beckoned. My pace had slowed to that of a crawl and my sweat was beginning to freeze against the inside of my clothing. I had covered around half a mile in an hour and I was physically and mentally exhausted. I could barely take a step without stumbling, jarring a knee, hip or ankle.

My pulk – the sledge tethered to my waist – was acting like an anchor on the ice, almost as if it had a mind of its own, slipping in one direction then another, knocking me off balance or dragging me backwards.

For the past few days I had been breaking trail through a

tortuous span of wind-blown waves of ice, known as sastrugi, that stretch across almost the whole of Antarctica. The abstract shapes are created during the winter months, when winds sweep across the continent at 120 mph. In places the surface appeared like an ocean, frozen in the midst of a powerful storm, where towering ice sculptures, some as big as cars, had risen up out of the snow. It was a spectacular sight but a nightmare to traverse, and the cause of many expedition failures. I had heard of people getting motion sickness as they tried to ski across sastrugi in a whiteout and I could now understand why. My sensory awareness had disappeared, and my balance was faltering with every slide forward. I began skiing with my legs further apart to try to balance better, and even spread my toes wide inside my boots in a forlorn effort to feel the terrain below me. My skis were constantly bending and flexing and I fell every few yards.

I fought hard to keep the negative thoughts out of my consciousness, the doubts that I was physically strong enough to finish. Instead I focused on the pain and the cold. It was a neverending mental battle not to lose my temper and curse the ice and the weather. On previous expeditions, as a member of a team, I had seen how tough, experienced men and women were mentally unravelled by sastrugi, striking the ice with their poles as they gave way to frustration. But when you are going solo you can't afford the luxury of anger.

I was being tested, I told myself. The Great White Queen was playing with me, seeing how far she could push me until I snapped, and when that happened it would be the beginning of the end. Once there was the tiniest chink in my mental armour, I sensed I would unravel quickly. An hour in those conditions is enough to break the spirit of any sane person, let alone three days.

Of all the physical challenges I've undertaken – including SAS selection – cutting a trail alone through a field of sastrugi, in zero visibility, on the way to the South Pole must count as one of the hardest.

After twenty-five years in the SAS, and having found myself in some tricky situations, I was confident I could manage my emotions better than some. I've always been able to think clearly, even when being shot at in the middle of a firefight, and I was able to bring that calmness with me to Antarctica. It was the one element of my character that I had always believed would never fail – but on this journey it was being tested like never before.

My knees, hip and ankles were bruised and raw, the sastrugi having tripped me more than thirty times that day. At one point my pulk slipped down a deep ridge and dragged me backwards, and for a split second I feared I would be pulled into a deep crevasse.

About an hour later I had a moment's loss of concentration, a slip – to be honest I can't remember – but the consequences were almost catastrophic.

I had just checked my compass, which hung from my chest on a hands-free plastic frame, for the umpteenth time, corrected my bearing, took a deep breath and pushed one ski forward when the ice beneath me disappeared. At the very moment I passed the point of no return I knew I was in trouble. I had stepped into a void and fell off an eight-foot-high ice ridge. I face-planted into the granite-hard ice below with such force that the impact drove the air from my lungs – it was as if I had been hit in the ribs by an uppercut from a heavyweight boxer. But worse was to come. My pulk followed and came crashing down on the back of my legs – the sucker punch.

The metallic taste of blood filled my mouth as the pain from the impact drilled up through my crumpled body. I assumed my legs were broken and my mind went into overdrive as I mentally rattled off a survival check list. I had to try to get to my satellite phone or emergency locator beacon before I passed out from the pain. But then rescue, I knew, would be delayed in a whiteout. And anyway, an aircraft couldn't land in the sastrugi field I was now in, so by the time any rescuers arrived I would be dead, entombed in ice, another soul lost to the Antarctic.

The pain of a fracture never arrived. My pride and chest were heavily bruised – nothing more. I took a few deep breaths and pushed the pulk away from my legs. I had dodged a bullet again, just like I had on a few occasions in Iraq and Afghanistan. I was once again reminded of my mortality, but there was no time to dwell. I had to push on and walk the fine line between caution and risk. If I slowed down and waited for the whiteout to pass I risked running out of food, but I was unlikely to walk away from another bad fall.

I took a few minutes to catch my breath and get my bearings. Every bit of my body ached and as I slowly moved forward again, I couldn't stop the doubts creeping in once more. How did I come to be here? How much more of this could I take?

1

BORN IN A BLIZZARD

Had we lived, I should have had a tale to tell
of the hardihood, endurance and courage of my
companions which would have stirred the heart of
every Englishman. These rough notes and our
dead bodies must tell the tale.

CAPTAIN ROBERT FALCON SCOTT

People always assume that I must like being cold. I don't like being cold – who does? But I've always been drawn to cold climates. My father says it was because I was born in a blizzard in February 1969, when much of the country was virtually encased in ice. According to family legend, my father was driving my heavily pregnant mother to hospital when the car got caught in a snowdrift and broke down. It seems that an al-fresco birth was only prevented when my dad managed to flag down a passing newspaper delivery van and hitch a lift to the hospital.

I grew up in Whaplode, a small, unremarkable village built on reclaimed fenland in Lincolnshire. For an aspiring adventurer, the search for excitement in that remote, flat and featureless land was long and often fruitless, although I did my best, wandering the fields, climbing trees, and building dens with my mates.

My family lived in a very modest but comfortable four-bedroomed semi-detached house. While we were by no means well off, I was happy enough, especially in my early years. My father, Philip, was a commercial air diver who worked as an underwater welder on the oil rigs – one of the more dangerous jobs in the world. He was often away from home for six weeks at a time, leaving my mum to look after me at the same time as working as a barmaid in the local pub. I was always so excited when he returned, happy to sit on his lap and listen to stories of how the vicious North Sea would endlessly batter the oil rig while he worked 'safely' below the water in complete peace and silence. But the joy of his return was always temporary, and I also knew that after a week or so the atmosphere would grow tense and family rows would begin. I suppose, looking back, it was the beginning of a period that would eventually end in my parents' separation and divorce.

My brother Nicky had arrived when I was six, and my parents' focus became directed towards him, which is understandable. Although my mum and dad were superficially happy, I sensed that there was always tension brewing just beneath the surface, and Nicky's arrival certainly added some stress to the family situation. Mum, Dad and particularly my auntie Christine (with whom I spent a lot of time) always said that I was an easy child, but Nicky was different. As a baby he was very demanding and quite a handful. Over the next couple of years, my dad came home less often and, when he did return, it was for shorter periods. By my ninth birthday my parents had separated. As Mum and I were accustomed to being on our own, this wasn't – as far as I remember – the traumatic event that it might have been for someone in a more close-knit family. My dad's absences were so frequent anyway that I simply took his long-term departure in my stride.

My dad moved first to the northeast of Scotland, to a fishing village called Stonehaven in Aberdeenshire. Sometime later, he and his new wife Val bought a derelict farm some 40 miles north of Stonehaven in a small rural village called Methlick. The farm was huge and consisted of a farmhouse, several barns, and dozens of acres of fields, woodland and pasture. Dad's plan was to rebuild the farmhouse and develop a clay-pigeon shooting lodge. It was no small undertaking, and every part of the complex was built from scratch. He was one of those people who was exceptionally gifted with his hands, and I never doubted for a second that he would eventually realize his dream. After much hard graft, Kingscliff Shooting Lodge opened its doors and became a mecca for shooting enthusiasts in and around Aberdeenshire.

Despite my parents' divorce and my lack of academic prowess, I managed to pass the eleven-plus exam and, in 1981, started at Spalding Grammar School. Like most eleven-year-old kids, I had no idea what I wanted to do when I became an adult, though I did have a growing, unquantifiable yearning for adventure. But three separate and unconnected events would eventually put me on a path that would finally lead me to Antarctica some twenty-eight years later.

Spalding was an academically demanding school and every pupil was expected to approach each subject with diligence and vigour. Unfortunately, though, I was one of those kids who was more interested in messing around than hitting the books. I was academically average at everything apart from maths, which was my nadir. No matter how hard I tried, I just couldn't get to grips with it. I must have been one of those kids who just managed to get across the line when taking the eleven-plus, and my inability to cope with algebra and geometry would dog me for the rest of my school days. I also struggled with – and had little interest in – the sciences.

One of my greatest loves was riding my BMX bike and learning new, risky tricks, such as a 180, a bar-spin, and the impossibly difficult tail-whip. These were tricks that required hours and hours of practice. BMX cycling and skateboarding are often dismissed as hobbies or sports, but for me they hold many life lessons. The skill gain is incremental and only comes with, quite literally, blood and sweat. I spent a lot of time falling off my bike, working out what went wrong and then attempting the same trick again. It was not a sport for the faint-hearted. You were going to get hurt, sometimes badly, and you had to learn to control your fear if you wanted to progress. Little did I know that the same principles would apply when I enlisted in the armed forces.

At around the same time I got into breakdancing. My mates and I formed a 'crew', as breakdancing gangs were known. On a Saturday we would head to Spalding, the nearest town, with a big square of lino and a ghetto blaster, one of those huge portable radio and cassette players. Various crews would arrive and a series of breakdance battles would begin, involving moves such as the windmill, where your body would spin around on the floor and your legs would flail around above your head. Other routines would combine the worm, head-spins and body-popping. I didn't appreciate it at the time, but you had to be pretty fit and flexible to pull off a decent series of dance moves.

Different crews would stand either side of the lino, doing their best to look mean and moody, and one by one we'd display our moves, trying to intimidate the opposition with various patterns and dances we had spent the previous evenings practising. The uniform consisted of a Nike cagoule, Sergio Tacchini tracksuit bottoms, and the most expensive trainers our mums could afford. It was a big deal for me at the time and helped me to define my individuality. The music was all American hip-hop from the Bronx, and included bands like Grandmaster Flash and the Furious Five, and Malcolm McLaren. My entrée into hip-hop and breakdancing came after I saw the 1984 cult movie *Beat Street*, which was like a breakdance version of *Saturday Night Fever*.

Like most teenagers, I looked forward to the summer holidays. My brother and I would be packed off by my mum to spend up to three weeks with my dad. Kingscliff was hidden amongst the rugged hills and valleys of Aberdeenshire. Woods and rivers defined the countryside, and herds of deer wandered freely. Eagles and birds of prey would soar overhead in the search for food, and for Nicky and I it was a limitless adventure playground, wholly different to Lincolnshire.

Dad had an old Vespa moped and we were allowed to zoom around the farm on it. My father also gave me a quick lesson on the finer points of tractor driving, and I met my first girlfriend, Alison, who – slightly bizarrely – was my stepmother's younger sister, but was the same age as me, twelve, at the time. While Dad was busy running the clay-pigeon shooting, Nicky and I were left to our own devices. The only rule my dad imposed was that we shouldn't get hurt – but in practice there was a certain amount of latitude in that. In fact, I had my first near-death experience there when I was thirteen. I'd been pestering my dad for ages to allow me to have a go at scuba-diving. He finally relented and took me to the picturesque harbour of Stonehaven for the day. With his usual lackadaisical child-rearing approach, my dad unloaded the scuba kit from the back of his car, gave me a five-minute briefing listing some dos and don'ts, and then said, 'Over to you, son.' He helped squeeze me into a wetsuit, placed the cylinder on my back and sent me on my way.

Suddenly I was in a new and fascinating world and I was mesmerized. Even in the murky harbour, there were fish of different sizes, large palm-like leaves of green translucent seaweed gently swaying in the current, and all sorts of maritime paraphernalia decorating the sea bed.

I had been down for around twenty minutes when I heard a loud, high-pitched noise almost like a drill. Curious, I decided to see what was happening. I'd just started to surface when a boat skimmed over the top of my head, so close that I could feel the turbulent aerated water from the prop bouncing on the front of my mask. I saw the rigid-hulled inflatable boat just a few feet ahead, in the process of zipping around the harbour. The occupants were staring back at me, a look of terror on their faces.

Dad, by contrast, was relaxing in the sun, licking an ice-cream, oblivious to the potential drama that had just unfolded a few feet away. Health and safety wasn't really a concept he grasped. It wasn't until years later, when I witnessed at close hand an accident also involving the propeller of a boat, that I realized how close I had actually come to being killed.

All lives are shaped by events in childhood, and probably one of the most important in mine came when a friend and I were sent to the headmaster after messing around during a maths class. Anyone sent to the headmaster by a teacher at Spalding Grammar knew that two things would happen. There would follow a lecture on the error of your ways for around ten minutes, then you'd be caned – a new experience for me.

My friend, Dave Marler, knocked on the door and went in first. I remained in the reception area, wondering whether to stand, sit or make a run for it.

I could hear the headmaster, Mr John Fordham, known as 'Skids' due to his middle name being Skidmore, lambasting him, pointing out why this misdemeanour would require the most severe of punishments. I swallowed hard and, in an attempt at distraction, I began to browse a bookcase in the corner of the headmaster's reception room.

One title jumped out at me. It was the small Ladybird book about Captain Scott. I sat down and began to leaf through the heavily thumbed pages, which recalled how Scott and his team of explorers had arrived in Antarctica in 1910 with the aim of being the first to reach the South Pole. The more I read, the more enthralled I became, and the less I cared for what the headmaster was planning. I read about how Scott set off, only to discover that he had been beaten to the Pole by the Norwegian explorer Roald Amundsen. Then came the return journey, with Scott's team battling against the elements, starvation and gangrene, and Captain Oates, close to death, sacrificing his life in the forlorn hope that the rest of Scott's party might survive, with the immortal words, 'I'm just going outside, I may be some time.' There was a picture of Oates disappearing into a blizzard, never to be seen again.

As I closed the book I could hear the swish and thwack of the cane landing on my friend's behind. But, empowered by reading about Scott and Oates – men who had faced death in the most horrendous of circumstances, I realized I did not feel in the least bit scared. In comparison to their fate, three whacks of the headmaster's cane were nothing. That night, as I studied the red, oblong welts on my backside in my bedroom mirror, I made a vow to myself that I would one day travel to the South Pole. Even then I knew it was more than a schoolboy fantasy – I felt almost as if it were my calling.

Unfortunately, I failed to learn from my mistakes and found myself in front of the headmaster again a couple of years later when Dave and I hatched a plan to break into the school one weekend for a bit of a laugh. During a maths lesson on a Friday afternoon, I deliberately left a window unlocked, and the following morning we climbed back in and began to wander around the school, which now seemed very different given that we were the only boys present. We went to the art department and found some Indian ink and painted tattoos on ourselves, then mooched around some classrooms, before I turned a fire extinguisher on Dave. Without any thought of the consequences, and completely absorbed by the moment, Dave and I began a massive fire-extinguisher battle, which only ended when we had used almost every canister in the school. As we surveyed the corridors, filled with foam and powder, the gravity of our actions began to dawn on us. We decided to make a run for it, but not before Dave had grabbed an expensive leather football from one of the classrooms.

On the following Monday, in school assembly, Mr Fordham unleashed his fury when he explained that vandals had broken in and caused considerable damage and that the police would soon be arriving to conduct an investigation. To say that I was terrified was an understatement. While I sat there squirming in my seat, Dave was a lot more chilled, but our undoing came when Dave began bragging about stealing the football. Within an hour or so we were summoned to the headmaster's office. I was first in and when presented with the evidence immediately coughed to the crime. Dave was in next and somehow was blamed as the ringleader. While I was caned, Dave was suspended for four weeks. The incident was a wake-up call, though. I knew that if I continued messing around I faced being expelled without any qualifications, so I decided to knuckle down, keep out of trouble, and try to salvage something out of the final couple of years.

By this point in my life, my mum had married a local man called George Smith, who had served for around six years in the Royal Marines. He had plenty of stories about travelling around the world, the tough training, and his pride in being a Royal Marines commando. His No. 2 uniform was hanging in my mum's wardrobe, and I used to imagine what I would look like wearing it. Various bits of military kit, such as webbing pouches, backpacks and the famous green beret were often around the house, and they helped to fire my imagination about what life would be like as a Marine commando.

One evening in April 1982, as I was eating my tea and watching the *Six O'Clock News*, I learnt that Argentina had invaded the Falkland Islands. Like much of the population at the time, I was confused. Surely the Falkland Islands were somewhere off the coast of Scotland was my initial thought, but I was quickly assured by the BBC that the islands were in the South Atlantic, some 8,000 miles away.

My stepfather came home later that night, furious that British sovereign territory had been sullied by the 'Argies', as he called them. He became even angrier when footage emerged a few days later of Royal Marines surrendering to Argentinian commandos.

Within a matter of weeks, a Royal Navy Task Force had set sail for the South Atlantic and a battle to retake the Falklands now seemed certain. The TV news was now nightly viewing in our household, with added commentary from my stepfather, who would frequently correct the reporters. He assured us all that now the Marines were involved, victory was a certainty. I was completely captivated by the conflict, not so much by the fighting, but by the sheer romantic adventurism of the entire campaign. The Royal Marines were in my eyes the epitome of heroes, and I vowed, although secretly, to join them.

Within a few months of the Argentinian surrender, a weekly fourteen-part history of the Falklands War was published. On one particular week, the magazine focused on the Royal Marines and the role they played. I don't know if the editor was on a secret mission to boost recruitment, but I failed to see how any teenage male wouldn't want to be part of such an elite fighting unit after reading about the heroic exploits of the Marines during the Battle for Mount Harriet and the Battle of Two Sisters. The crowning glory for me, though, was a picture of a battle-hardened Royal Marine, yomping across the Falklands with a Union Jack tied to his radio aerial, flapping in the South Atlantic wind.

The photo took pride of place on my bedroom wall, and I went to sleep every night imagining myself in the Marines, with nothing but a lifetime of adventure to look forward to. As the months passed, I began to tell my family, friends and even my teachers that I had no interest in A-levels or university; instead I would join the Royal Marines at sixteen. My enthusiasm for a military career was often greeted with undisguised scepticism by most of my teachers, who would ask whether 'I was sure that's what I wanted', or advise me that it was 'extremely tough and very few are selected' and I should 'not put all my eggs in one basket'. But I remained undeterred. I soon learnt that I thrived on negativity; the more people advised me against a career in the armed forces, the more determined I became to join up.

Two years later, just before the start of the summer holidays in 1984, I was further inspired by a book by Sir Ranulph Fiennes titled *A Talent for Trouble*. Here was a man, I concluded, who grabbed life and ran with it. He walked the walk, and cared nothing for what others thought. I decided I would do the same, and straight away I began to plan my own expedition. With just a week left of school before the summer break, I decided I was going to cycle to my dad's house in Scotland, some 500 miles away. As with previous summer holidays, my parents had already decided that Nicky and I would spend the first three weeks with my dad in Aberdeenshire. Normally we travelled up to Scotland by train on our own, but this year I wanted to test myself.

Even then, as a fourteen-year-old, I was aware that simplicity was key. The fewer moving parts there were in my plan, the less likely it was to go wrong. I took an old and beaten-up road map from my stepdad's car and began preparing my bike, a three-gear racer with drop handlebars. I tightened the brakes and pedals, greased up the chain, and invested in a very modest puncture-repair kit. I planned a route that kept me on B-roads, away from motorways and busier trunk roads, passing through villages where I could buy food, and wooded areas where I could sleep. Above all else, I was utterly convinced that the trip was achievable and was a goer – providing my mum agreed.

Arriving home on the last day of school, I went to find her.

'You know we're going to Dad's next week,' I said cautiously.

'Yes', she replied as she started laying the table for dinner for me and Nicky.

'Well, I want to cycle.'

My mum stopped what she was doing and looked at me disbelievingly. Her brow furrowed and her ever-present smile fell away from her face. She went to speak but I was too quick.

'It will be fine, Mum. I've worked out the route; I've got a map and saved some pocket money. I will take some food with me and I have a sleeping bag. I've checked the weather forecast and there is no sign of any rain. I really, really want to do this. I'm fourteen, I'm old enough, and in two years' time I will be a Royal Marine, so this is nothing. You have to let me go,' I insisted.

'OK,' she added, seemingly accepting that any form of protest would be useless. 'If that's what you really want to do.'

'It is,' I interjected quickly.

'OK. Well, I'd better tell your dad.'

'No. Please don't. I want to surprise him.'

My mum smiled. 'OK then. Nicky will have to go up by train, though. There's no way he can ride a bike that far.'

That night I began to make a note of what I thought I would need for the trip – a change of clothes, a very thin, lightweight sleeping bag, food for the first day or so, and some pocket money. I didn't own a tent and, as it was summer, I decided that I would sleep out in the open – a new experience for me. The only water I planned to carry was in a bottle attached to the frame of my bike. I figured I could top it up from streams and public toilets.

Two days later, following a large breakfast, I said my goodbyes and set off on my very first adventure. My excitement powered my legs for the first few hours before I stopped for a drink and a map check. My plan was to cycle until it was almost dark – I had no lights on my bike – then find somewhere to sleep.

On the first day I managed to cover almost 90 miles, passing through numerous towns and villages. My plan was to make it to the Humber Estuary, if possible, and that first night I slept exhausted on the northern side of the Humber River, beneath the bridge, on the outskirts of Hull. I woke the following morning as the sun rose, bringing warmth to my chilled body. It was the first time I had ever witnessed a rising dawn, and I viewed it as a sort of good omen, as if nature was willing me on. I was soon on my way again, heading north and feeling completely free for the first time in my life. I felt as if I had unshackled myself from the rules of normal society.

I travelled further north day by day, eventually crossing into Scotland, and watched as the landscape changed. Cycling connected me to nature in a way that other forms of travel didn't. I slept in woods, washed in streams, and bought food from small village shops where I was asked what I was doing so far away from home. My answer was often met with incredulity and I felt sure that I would be reported to the police as a runaway if I stayed in one place for too long.

By the fifth day, I had reached Aberdeenshire, and was starting to feel extremely tired, but I managed to reach my father's isolated farm by teatime.

I knocked and opened the door of his farmhouse, and walked into the kitchen where he was having a cup of tea with his wife, Val.

'Louis,' he said with a look of surprise and confusion. 'What are you doing here? I thought you were due in a couple of days' time?'

'I was,' I said, adding, 'but I decided to cycle.'

Now there was more confusion.

'From the station? You cycled from Aberdeen Station?'

'No from home, from Whaplode. I've spent the last five days cycling. I've been planning it for ages. I wanted to go on an expedition, so I planned my own.'

'You're having me on – that's over 500 miles,' he said, still doubtful.

'Actually it's 515, Dad. I managed around 100 miles a day – sometimes more, but the last part was very hard.'

My dad still refused to believe me and began laughing.

'I'm going to phone your mum,' he said, heading out of the kitchen.

Five minutes later he returned. He stood in front of me, hands on hips, with a huge smile across his face. 'Good for you,' he said, now finally convinced that I was telling the truth. 'You must be exhausted and hungry.'

I spent the rest of the evening telling my dad about my adventure, and the look of admiration on his face was all I could have hoped for.

The following evening Nicky arrived, and we spent the next three weeks climbing hills, exploring remote valleys, going on long walks, building dens, tracking deer, trying and failing to catch brown trout with our hands. It was a blissful summer. By the time I returned home three weeks later, by train this time, I was a different person – more confident and self-assured.

The next two years of my school career seemed to race by. In early 1985 I passed five O-levels in English Language, Literature, History, French and Religious Studies, and while my teachers talked about further education and university, my unwavering ambition was still to join the Royal Marines. It had become the single focus in my life and was something of a joke amongst my friends. When there was talk of what we were going to do when we 'grew up', it was quickly pointed out that I was going to become a commando. I believed that I was going to join the Marines with the same certainty that I knew the sun would rise every morning.

2

REJECTION AND ACCEPTANCE

Success is not final, failure is not fatal:
it is the courage to continue that counts.

Winston Churchill

On the Monday after I left Spalding Grammar School, my mum
and I travelled to Peterborough for an interview in the Royal
Marines Careers Office, which was located on the bustling high
street. Mum made sure that I looked suitably smart, wore a tie
and combed my hair. If she had any doubts about my career
plans, she kept them to herself. I couldn't remember when I'd felt
more excited.

I was sixteen years old and nearly six foot but I weighed
around 70 kilograms soaking wet and had the physical bearing of
a beanpole. I had excelled at school in running, and had won the
school cross-country championship in my final year, something I
have always remembered with pride, but I was about to find out
that having a lean runner's build came with a price.

A sergeant, whose sheer physical presence seemed to fill the
entire room, ran the Royal Marines recruiting office. He welcomed us both with a smile and a crushing handshake, which
forced me to stifle a grimace. His uniform was immaculate. His
black commando boots had been polished to a mirror finish. His
olive-green trousers had razor-like creases and his green military
jumper boasted the red Royal Marines commando shoulder flash.
He looked strong, fit and aggressive, and boasted a chest that was

big enough to belong to a comic book superhero. I convinced myself that he had probably dispatched many Argentinian soldiers during the Falklands War. He was everything I wanted to be.

The office walls were covered with posters promising aspirant Royal Marine recruits a life of challenge, adventure and travel. I felt as if I had finally found something I had been waiting my whole life to discover.

The recruiting sergeant ushered Mum and me towards a desk and then began a brief interrogation.

'Why do you want to join my Corps?' he asked suspiciously, as if I had committed some sort of offence.

'Because the Royal Marines are the best fighting force in the world. The best trained, the fittest. They can deploy by air, land and sea, and I want to be part of that.'

'Are you fit? What sports do you do?' he enquired.

'I play all sports but I'm also a pretty good cross-country runner.'

'How do you know the Royal Marines will be for you?'

'Well,' I responded, knowing this was a critical question, 'I've read a lot about the Royal Marines, I followed what they did in the Falklands War and my stepdad served in the Royal Marines for five or six years.'

The last part of my answer drew the smallest of smiles on his otherwise inscrutable face. The sergeant then began explaining the lengthy and complex process of joining the Marines. There would be an interview, academic tests, and then a medical, and if I passed those I would have to attend and pass a four-day Potential Recruits Course (PRC),* and only then would I be offered a place at the prestigious Commando Training Centre Royal Marines (CTCRM) at Lympstone in Devon. In all, the sergeant explained, joining could take up to a year. He must have seen my disappointment, because he promised that if I passed the tests

---

* Renamed Potential Royal Marines Course (PRMC) in the 1990s.

and medical, he would do his best to get me on a recruits course as soon as possible.

Two weeks later, I returned once again to complete the interview and the test, both of which I sailed through. That afternoon I was sent to the local hospital for a medical, where I was examined by a doctor who poked, prodded, frowned and smiled, but didn't give me any clue about his verdict. He filled in a form, sealed it in an envelope, and told me that I should now return to the careers office.

I had assumed that this was a formality. I was fit, I had cycled to Scotland, I had won the school cross-country race and I could breakdance like a demon. But it all mattered little. The recruiting sergeant opened the envelope, removed the form, and bluntly stated that I had failed the medical.

I was lost for words. I thought that somehow I was dreaming and that I would wake up soon. I had spent the last few years telling anyone who was willing to listen that I was going to become a Royal Marine. I had left school, ceremoniously burning my books and school uniform. This could not be happening. The recruiting sergeant must have sensed my disbelief, so to reinforce what he had just told me, he pointed at the form.

'Look,' he said unsympathetically. 'The doctor says your body is not ready for the rigours of commando training. Physically, you are still developing. You are just sixteen. If you went now you wouldn't last the course. You could be injured, badly, and medically discharged. Your career would be over before you started. Believe me, it happens.'

I nodded but said nothing as the disappointment sank in.

'You need to bulk up and put on some weight. If you are still interested in six months' time, then come back and you can have another medical. If you pass, I'll put you on the next PRC.'

With that he stood up and shook my hand, clearly not wishing to waste any more time on someone who couldn't get over the first hurdle.

I drove home with Mum in complete silence.

Mum tried to be cheery, but her kind words fell on deaf ears. For the first time in my life I had experienced real failure, and it left me feeling utterly crushed. In a matter of minutes my whole future had disappeared.

By the time I had arrived home, I had resolved to try again. I would get stronger, I told myself, I would eat loads, I would demand steaks for every meal, I'd go to the gym, I wouldn't be beaten. But first I'd have to suffer the indignity of telling my mates that I had failed the medical and live with the ignominy of being too scrawny to be a Royal Marine.

My stepdad, George, was brilliant. He told me not to worry, and explained how many of his friends had failed the medical and then gone on to achieve great things in life. He also insisted that we enrolled in the local gym together, so that we could both work out. He promised that he would help prepare me for the various tests I was likely to face when I eventually passed the medical.

With six months to kill before I could try again, my mum did what mothers do and found me a job at a local engineering firm in Whaplode, called Greens. It was a factory making farm sheds and metal building materials, and my job was to help service the steel cutting machines. The place seemed to be full of miserable, middle-aged men who had done little with their lives; my arrival offered them the perfect target for their inner self-loathing. I was the butt of jokes and pranks. I was expected to make tea whenever they wanted a cup, and to do the jobs they felt were beneath them. Unsurprisingly, when they discovered that I was planning to join the Royal Marines, they couldn't wait to question whether I was made of the right stuff. But once again, I used their negativity to fire myself up, and became even more determined to pass the medical. Within two weeks I had regained my focus and set about planning what I needed to do. When my stepdad couldn't take me to the gym, I would do press-ups and sit-ups in my bedroom. I went for runs and immersed myself in Royal Marines history.

Six months later, in late 1985, I was back being examined by the same doctor, who remembered me and said that he was amazed by the transformation. He told me that I had passed and I virtually ran back to the recruiting office and pleaded to be sent on the next available PRC. The unsmiling and unimpressed recruiting sergeant told me that I would get a date through the post with details of the course.

For the next six weeks I anxiously awaited the arrival of the morning post every day, as I gorged myself on an oversized breakfast – all part of my weight-gain programme – until eventually, a letter arrived telling me to report to the CTCRM.

A few weeks later, my mum and stepfather drove me to Peterborough Station, where I boarded a train first to London and then on to Lympstone Commando – the name of the station that serves CTCRM. The training centre is situated in South Devon, next to the River Exe and close to Woodbury Common. The surrounding country is rugged and criss-crossed with rivers, streams and woodland, and is ideally situated for turning civilians into Royal Marines.

By the time the train arrived in Lympstone, it had picked up around thirty potential recruits of all ages and backgrounds. A few chatted but most of us were quiet, lost in thought and wondering what the next four days had in store for us.

A waiting drill instructor greeted us on the platform, took a register of names and informed us all that the course had now begun.

'I will tell you once and once only what to do,' he barked. 'Pay attention at all times. Listen to what is being said to you. If you don't understand, put your hand up and ask a question. The Royal Marines will not lower its standards for you. If you all meet the grade you will all pass. If you don't you will fail. Now follow me.'

Although the corporal was only walking, his pace was frenetic and most of us were jogging to keep up. By the time we had arrived at our accommodation, we had become a group of sweaty stragglers.

Our beds that night were in an empty barrack block, but I can't recall sleeping. The following morning we were issued with a set of overalls, and then given a briefing by the chief instructor.

'We are looking for people who have potential and display the commando spirit to complete commando training,' the drill instructor said, looking each of us in the eye and emphasizing the word 'potential'. 'You will be given a series of tests, not all physical, and you will be scored on each one. Some of the tests have a minimum standard attached to them. If you just pass at the minimum level, you will receive one point. Do not aim to do the minimum, always aim to do your best,' he continued.

'We will look at your ability to work as a team, absorb new information and understand simple orders. You will have the opportunity to ask plenty of questions. This is your opportunity to look at us and ask, *Is this for me?* We will also be doing the same.'

Over the three remaining days I took part in runs, crawled through flooded tunnels, climbed ropes, and attempted various elements of the notorious Royal Marines assault course. One of the tests was a three-mile run, called a Basic Fitness Test (BFT) and I came second, which gave me a massive boost, especially when one of the PTIs asked what my name was and said, 'Well done.'

There were one-to-one interviews, some memory tests, and we were quizzed on our knowledge of the Corps and why we wanted to join. There were basic lessons on how to strip and put together a 9 mm sub-machine gun and a 7.62 mm self-loading rifle. The various instructors would name several parts of the weapon and then ask each one of us what they were. The assessment was continuous and unrelenting. Then, in a flash, it was all over, and on the final day we stood in a line outside the barrack block that had been our home, waiting to hear our fate.

The corporal read out a list of names and we were split into three groups. Those who had taken and failed the course before had warned us of this final, nerve-wracking stage. What was known was that amongst the three groups one consisted of outright failures, another were those candidates who had failed but would be invited back to retake the course at some time in the future, and the third were the few who had passed.

I looked at my group and had no inkling which one it was. Each group was marched into a classroom and then departed from another door. I was in the last group, and by now I felt as if I was having palpitations.

The first thing I noticed when my group entered the classroom was the unsmiling face of the corporal who had watched us like a hawk since our arrival four days earlier. I feared the worst. My palms were sweaty and I began wondering what I would tell my friends and family if I had failed. For the past few years, all I had wanted was to be a Marine, and now it was crunch time.

'Congratulations, gentlemen,' the drill instructor said. 'You have passed the PRC and you will be offered a place at CTCRM to commence training in the near future. You have shown us that you have the potential determination and character to become a Royal Marine commando. But this is just the first step. You now need to go home and prepare for the hardest thirty-two weeks of your life. Good luck.'

I was utterly elated and remained on a high all the way home. My mum and stepdad had arranged to meet me at the station, and they were both close to tears when I told them that I had passed.

A few weeks later, on 21 October 1985, I closed the door on my childhood as I walked through the gates of CTCRM in Lympstone, Devon. A new life beckoned and I couldn't wait for it to begin.

3

COMMANDO SPIRIT

If it's not raining, it's not training.

Unknown

As I walked through the gates of CTCRM, it dawned on me that I was a boy in a man's world, and I had to fight with all my might to stop my legs turning to jelly. I was the youngest of an intake of around thirty raw and nervous recruits, aged between sixteen and twenty, about to embark on thirty-two weeks of the hardest basic training in any of the world's armed forces. We were all put into 297 Troop, a junior troop. Lympstone has a troop intake every two weeks; there was an adult troop two weeks ahead of us and another two weeks behind.

The Commando Training Centre resembled a small, perfectly neat concrete village. Everything had its place and purpose, including its inhabitants. No thought whatsoever had been given to making it look welcoming. The base was a study in military discipline. The entire complex was spotless; every building looked as if it had recently been painted. Anything made of brass was polished to a mirror finish, and in the distance someone, somewhere always seemed to be shouting. The training centre was composed of classrooms, a vast cookhouse, a chapel, a huge gym, ranges, accommodation blocks where the recruits slept and even a swimming pool. The two most dominating features were a large parade square – the domain of drill instructors and best avoided – and the 'Bottom Field', where the feared Assault and

Tarzan courses were located. It was effectively a factory with one job: to turn civilians into Royal Marines within thirty-two weeks.

The Recruit Training Troops – a troop is the Royal Marine term for a platoon – were all at various stages of training, from the newly enlisted, like us, to the King's Squad, who had earned their green berets and were a week away from becoming the real deal.

The first two weeks, known as Foundation, were a whirlwind of kit supplies, introductions, briefs and camp orientation. Everyone swore an oath of allegiance to the Queen and we were issued with our service numbers – mine was to be P044868H. Our heads were shorn into a close crop and we were 'taught' how to wash, shave, shower, clean our teeth and wipe our backsides. Detailed and probing medical examinations were conducted to ensure we had no infectious diseases and that everything was where it was supposed to be. There were ironing lessons in the evening, followed by hours spent learning how to make our bed packs to inch-perfect measurements.

At the end of the Foundation Phase, the troop was marched over to our new barrack block, where it became immediately apparent that our lives were about to get a lot harder. The accommodation was functional and austere. Each room contained eight beds and no privacy. A single wooden wardrobe was provided for military-issued equipment: uniform, webbing, boots and helmet, and a footlocker for any personal equipment, such as civilian clothes.

Later that first evening as we sat on our unmade beds, some no doubt having second thoughts, our section corporal arrived and introduced himself. He was a Falklands War veteran and had the face of a man who looked as if he had bypassed his childhood. He was devoid of any form of humour, and in the entire thirty-two weeks of basic training, I don't think I ever saw him smile. He explained he was only ever to be addressed as 'Corporal', nothing more.

He told us he would teach us everything we needed to know to fight and survive on the battlefield, and woe betide anyone who didn't meet his exacting standards. His life, he explained, was the Corps, and he was determined only to send competent, capable Royal Marines to commando units. He would, he told us in no uncertain terms, fail us all if we didn't make the grade. His words of command were issued in a series of growls, and he appeared to have nothing but complete contempt for us all. My home, family and all of my creature comforts now seemed a long way away.

The corporal also explained that we were now not only responsible for keeping ourselves clean, but also the toilets, showers, sinks and our sleeping quarters, which he would inspect every morning at 7 a.m.

It soon became apparent that – no matter how hard we tried – nothing was ever good enough. He would find dirt everywhere, in places where we had spent hours cleaning. The days were long and began around 5.30 a.m. and ended just before midnight. We were constantly tired and referred to as Nods by the instructors, due to our habit of nodding off during lessons.

Within a few weeks of arriving at CTCRM, I knew with absolute certainty that I was never going to be any good at drill. I just didn't possess the level of coordination needed to march in time with everyone else. I used every ounce of both mental determination and concentration I possessed but, within a dozen paces of the order 'Quick March!' being given, I was somehow out of step. The harder I tried, the worse my coordination became.

There was no place to hide during drill sessions. Any mistake, any hesitation, became immediately identifiable by the corporal watching our every move. Drill was created to move large groups of men around the battlefield in time while obeying words of command without question. In the modern armed forces, its function was ceremonial, but the concept of teamwork and the unquestioning obedience to orders remained relevant. Drill was, and still is, taken incredibly seriously, even in the Royal Marines. While other recruits feared the long exhausting runs, combined with the assault courses of the Bottom Field and hardcore gym training sessions known as 'beastings', I myself had nightmares about drill.

Every waking minute during every day of recruit training was timetabled, and the pace of life was unrelenting. Most mornings began with a room inspection followed by a run, often through muddy rivers in the surrounding countryside of Woodbury Common, or exhausting PT sessions in the gym that regularly ended with a 100-foot rope climb.

After PT sessions, we were expected to conduct a quick change into immaculate uniforms ready for a lesson in basic tactics, weapon training, map reading or military history. We learnt how to yomp – marching cross-country for what seemed like endless miles while carrying weights of up to 18 kilograms on our backs – and how to bayonet the enemy. Every phase of the military syllabus was examined and tested, often in the form of an exercise on Dartmoor, in deteriorating weather conditions as winter approached.

Meals lasted minutes and were wolfed down furiously, so that we could prepare for the next lesson or training task. Only Sunday afternoons offered us any form of respite from the unrelenting demands of our transition from civilian to Royal Marine. As well as everything else, there was the Royal Marine lingo, which we were expected to master. Food was 'scran', hot drinks were 'wets', stories were 'dits', the kitchen was a 'galley', toilets were 'heads' and Army soldiers were 'pongos' (because everywhere the Army goes, the pong goes).

The pace of life was furious and I, like everyone else, was on the back foot most of the time. As the days merged into weeks, the marches became longer and the weight we carried grew heavier. To be honest, I was struggling with the course. It was a major culture shock. There were never enough hours in the day to get our kit prepared, clean our boots and absorb all of the new skills we were expected to acquire. I wasn't alone; everyone found it tough and, looking back, it seems clear the course was designed to put us under pressure and see how we reacted. Despite the physical and mental challenges, though, I never for a moment doubted that a career in the Royal Marines was what I wanted.

It wasn't all pain and suffering, and I soon discovered that I was more at home in the field than the barrack room. Fieldcraft is the skill of being able to operate stealthily without being seen by the enemy. The instructors taught us about camouflage and concealment, how to move at night, silently using natural cover. It came almost naturally to me.

We learnt how to take part in recce and fighting patrols, ambushes, and how to identify targets. Crucially, our instructors also taught us how to maintain ourselves in the field by keeping dry when it was raining, looking after our often-soaked feet, cooking and feeding ourselves using military rations, and how to secure our positions at night. It was what being a Royal Marine was all about.

Christmas came and went, as did my seventeenth birthday in February, which for the first and only time in my life I completely forgot about until a few days afterwards. As the training became harder and more intense, the junior troop began to decrease in size. Some of the recruits simply gave up, admitting that the Royal Marines wasn't for them. Others sustained injuries, and a few were told that they needed to retake a section of the course and were sent back to join another troop at an earlier stage of training. The one constant throughout the thirty-two weeks of training, however, was our corporal's unwavering antipathy towards his section. He seemed to enjoy making life as miserable as possible, and we were convinced that he regarded recruit training as a punishment posting.

His reference point was always the Falklands War. Mess up on the drill square and he'd unleash volley after volley of abuse upon us all, and would question how we could ever hope to survive combat when we couldn't follow basic instructions.

The corporal took a particular dislike to me. I can't be sure why, but I think it came down to my relative immaturity. I wasn't cocky or mouthy; I learnt very quickly that the trick to making it through basic training was to follow the line of least resistance, which for me meant trying as hard as possible and staying out of trouble. But for some inexplicable reason, my corporal resented this.

Easter arrived, and with it the luxury of two weeks' leave. A corporal from another section marched us down to the station and, just as we boarded the train, I made the mistake of telling him how much I was looking forward to a couple of weeks away from our own section corporal.

Like many of my fellow recruits, I slept for most of Easter leave, and ate as much as I could of my mum's home cooking, which I had missed terribly. My stepdad wanted to know whether much had changed since his recruit training, and we came to the conclusion that it hadn't. Then, in a flash, my leave was over, and I was back on the platform at Lympstone Commando. I was just about to head back into camp when I heard the unmistakable voice of my section corporal.

'Rudd!' His scream echoed along the platform. 'Get your arse over here.'

I felt as though I was locked in the sights of a sniper. As I ran towards him, two thoughts occurred to me: how did he know which train I was on, and what had I done wrong? In the ten or so seconds it took me to run down the length of the platform, I had the answers. Firstly, someone had told him about my comments just as we were going on leave, and secondly he must have been waiting for every train to arrive, brooding over what sort of punishment he was going to exact.

'Thought I wouldn't hear, didn't you, Rudd?' he said, his anger spitting into my face.

'Corporal,' was my only response, and was an affirmation of my wrongdoing.

'Taking the piss out of me for a few laughs?' he added.

'Corporal,' I offered meekly.

'Meet me on the Bottom Field in ten minutes in your PT kit.'

I turned and shot off, running up the hill through the gates and into my room. Some of the other recruits were already back, relaxing, reading and chatting. I dropped my bag, ditched my clothes and changed into my PT kit. 'I'll explain later,' I said, running out of the door.

The sun was beginning to set and the wet dew was already beginning to freeze into a thin layer of ice by the time I met up with my section corporal. The comfort of my home in Whaplode was still fresh in my memory, but it seemed a long way away as I began carrying out press-ups, sit-ups and star jumps in the mud and cold. He and I both knew that this wasn't strictly a formal punishment. But he was also aware that I would not make any formal complaint. My only choice was to take what he dished out and show no sign of weakness.

After half an hour, he ordered me to run over to the water tank. In usual conditions, troops would crawl across the obstacle on a rope suspended above.

'Get inside, Rudd,' he said, snarling.

I didn't blink or make a sound as the freezing water sent my breathing into a spasm.

'Now sit down,' he added.

The water reached up to my neck and I was seriously cold. After five minutes or so he relented, possibly realizing he was taking the ad-hoc punishment too far, even for him. My arms and legs were stiff with cold and I struggled to run back to the barrack block. My roommates were furious after I had explained what had happened.

Later that evening, the troop sergeant gave the usual evening brief, listing the events for the following day and the rest of the week. Various recruits and sections were assigned duties. He also warned us to snap back into the Marine way of life as soon as possible. Then, almost as an afterthought, he announced: 'Recruit Wright won't be returning. He was killed in a car accident. Right, that's it. Get back to your rooms and prepare for tomorrow.' It was only then that I noticed the halo that had been drawn above Wright's head on the troop photograph hanging outside the instructor's office: classic Marines dark humour.

I was stunned. Wrighty had been one of my best friends in recruit training. The recruits exchanged shocked looks with one another. At first I thought the sergeant was being callous, but then over the next few days accepted that he was probably trying to be kind. Recruit training is not the place to dwell on someone's death. It is hard enough without questioning the meaning of life. It was the first – although not the last – time that I would lose a good friend while serving in the armed forces.

I went to bed that night thinking of Wrighty and the fun we'd had. I realized that life was short, precious, and above all should be cherished.

As the course progressed, the recruits were given a little bit more time off and were allowed to leave the camp, especially during the weekend. One Saturday night, at around the twenty-week point, me and a group of lads ventured into Exeter for a night out. A few of us ended up in a grotty nightclub in the city centre called Winston's where, in a time-honoured way, we all got drunk. At the end of the evening, myself and another recruit called Rob went to catch the last train back to Lympstone, which is when we realized that we had spent all our money.

Undeterred, we boarded the train and breathed a sigh of relief that we didn't have a seven-mile walk ahead of us. Our plan, if we were asked for our tickets, was to appeal to the conductor's good nature and explain that we were recruits on a night out and that we were now penniless.

When the conductor appeared, it soon became apparent that he was not going to show us any sympathy at all. Instead he said he was going to get the train stopped at the next station and contact the police. Although drunk, I realized this wasn't funny. You can get away with quite a bit as a recruit, but once the police get involved, events can take a nasty turn. As the conductor headed back down the carriage, my friend and I hatched an ingenious plan. Rather than be captured by the police, we decided to hide. Unfortunately, the number of hiding places on a train are limited.

'What are we going to do?' Rob asked, realizing for the first time the severity of the situation.

'On the roof,' I said, pulling down a window. 'We're going to have to climb on the roof. It's the only way. Otherwise we're going to get caught and corporal will do his nut. We could even be back-trooped or booted out.'

The thought of him finding out that we had brought the name of his beloved Corps into disrepute was enough to motivate any recruit into a life-risking situation.

'Rob, just follow me. Hold on tight and keep your head down.'

I pulled the window down and climbed on to the roof of the train as it sped through the South Devon countryside. Both of us were gripping on to the roof with all our strength, having sobered up pretty quickly when faced with a potentially life-ending error of judgement. It was pitch black and raining, but we both had the good sense to keep our heads down as the train rattled along under bridges and tunnels with just a few inches' clearance. The train pulled into the next station where a transport policeman was waiting. As he boarded, we could hear the conductor ranting about us and wondering out loud how we had disappeared.

Eventually the train departed, and a few stops later arrived at Lympstone Commando. Rob and I slid off the roof onto the opposite side of the platform and waited for the train to depart. As it disappeared into the night, we looked at each other and burst out laughing. Both of us were covered in black soot, our clothes ruined. Surprisingly, no one questioned us as we showed our ID at the gate to the base and headed back to our barrack room, where we both fell asleep fully clothed. Fortunately, we had the Sunday to get ourselves sorted out before the section corporal appeared.

The Commando Tests came in week thirty. If we passed, we would be awarded our green berets. It was a pivotal moment in the course. By this time our troop had been bolstered with extra recruits, who had either been back-trooped or been injured in a previous attempt to secure the green beret. They came with foreboding tales of pain and suffering, but my section had bonded well over the previous months, and we vowed to work as a team to get each other through.

The challenges started on a Saturday with the Endurance Course, which was six miles in length and began with a two-mile run across the moorland of Woodbury Common. The course took us through flooded culverts, pipes and rivers. The last four miles were effectively a sprint back to camp, and the course culminated in a live-fire shooting test in which dropped shots added a time penalty.

The Sunday was a rest day, and then the week began with a nine-mile speed march. This was done whilst carrying our fighting order, which consisted of webbing, ammunition pouches, water bottle holder and kidney pouches containing mess tins, spare socks, and weighed around 16 kilograms in total. We also carried a rifle, which was slung over a shoulder. The march had to be completed in ninety minutes – anyone who failed was given just one more chance a few days later. The following day it was the Tarzan Assault Course, which was more to do with confidence than fitness, even though it had a time limit of thirteen minutes.

By far the hardest test was the thirty-miler. It mattered not how fit or strong you were, the thirty-miler was tough. The route took us across some of the most arduous terrain that Dartmoor has to offer, and had to be completed in eight hours. It was designed to hurt and make you work as a team. There were moments when I felt that I had nothing left to give. I had no intention of giving up, I was just unsure if I had the physical capability to put one foot in front of another, but with the help of my mates and the ability to push through the pain barrier – a skill my corporal had helped me acquire – I completed the march, and along with most of my section was awarded the green beret. While we all congratulated each other, our troop sergeant informed us that this was just the start of our training and that we had much to learn when we arrived at our units.

Two weeks later it was all over. In the space of thirty-two weeks I had transformed from being an insecure and immature teenager who hadn't needed to shave, to a trained Royal Marine commando. I had gained more than six kilos in weight and four inches round my chest. The Passing Out Parade on Friday 13 June 1986 was a real family day. My mum and stepdad were there, as were my father and his Scottish wife, Val, along with cousins, grandparents and my brother Nicky.

It was – and remains – one of the most important and proudest days of my life. I was posted to 40 Commando in Taunton, Somerset – my first-choice unit. I was still only seventeen years old, but I felt ready to take on the world.

4

BOOTNECK LIFE

Train hard, fight easy.

ALEXANDER SUVOROV

In March 1971, three off-duty British soldiers, John McCaig, who was aged seventeen, his eighteen-year-old brother Joseph and Dougald McCaughey, twenty-three, were all murdered by the IRA. The three, all members of the 1st Battalion, Royal Highland Fusiliers, were believed to have been lured to their deaths after drinking in a bar in Belfast city centre. The deaths led to a national outcry, due to their age and the fact that they were murdered while off-duty and dressed in civilian clothes. Up until that point, only three British soldiers had been killed in Northern Ireland, and those deaths had occurred during riots while the troops were on duty. The murders became known as the Honey Trap Killings because it was believed that the soldiers had been invited to a party by female members of the IRA. As a direct response to the killings, the government raised the minimum age to serve in Northern Ireland to eighteen.

I was unaware of this small but crucial fact when I arrived fresh-faced and over-eager for my first day's service at 40 Commando in Taunton, Somerset, in August 1986. The unit was halfway through a six-month tour of South Armagh, otherwise known as Bandit Country, and instead of being sent out as a re-inforcement, as the other recruits from 297 Troop had been, I was told that I was remaining in Taunton as part of the Rear Party.

It was possibly the least spectacular start to a military career anyone could wish for. The role of Rear Party was essentially to look after the barracks while the main force was away on operations, and it was composed of troops who were unfit, medically downgraded, or ill-disciplined – and me, the Fucking New Guy or FNG, who was too young to deploy.

My new unit, 4 Troop Bravo Company, 40 Commando, eventually returned to Taunton two months later, and once again I found myself at the bottom of the food chain. Bravo Company was full of old sweats, seasoned boot-necks, who had fought in the Falklands and taken part in the landings at San Carlos Bay. Many had multiple tours of Northern Ireland under their belt. In those early days I was tolerated more than accepted. I quickly became the 'brew bitch': the person responsible for making 'wets' for corporals, the troop sergeant and troop commander, a young public-school-educated lieutenant, and mostly anyone else who wanted one. I made no complaints. I kept my head down, worked hard on exercise and earned the reputation of a Marine who could be relied upon, even though I was inexperienced.

Life within a commando unit in those days was very different to recruit training at Lympstone. It lacked the intensity of those long, action-packed days where the recruits would be under the training teams 24/7, and where every minute of every day had been pre-planned. In a unit, whole days seemed to pass by without much being achieved.

Most mornings would begin with some form of physical training, but the rest of the day would fritter away with a bit of low-level tactical training. After spending six long months living with the stress of knowing that every time they left the security of their base they were potentially an IRA target, the guys in my troop welcomed a few weeks of doing relatively little. But for a young, impressionable Marine, desperate for action, my new life left me feeling a bit deflated. I had spent the hardest eight months of my life joining an elite fighting force, only to discover that the reality could be very different.

Fortunately, the exercise season soon kicked in, and the whole of 3 Commando Brigade – which included commando units plus all attached arms – took part in a major exercise across Europe lasting several weeks. For 40 Commando that meant undertaking a series of amphibious beach landings and assaults on enemy positions in Denmark and Germany, and as the FNG I loved every minute of it.

But the real highlight of that first year in the Royal Marines was six weeks of jungle warfare training in Brunei called Exercise Curry Trail. Word quickly went around the company that jungle training was the most demanding training imaginable, and I couldn't wait. This was what I had joined up for. Weeks of briefings and preparations followed; we were inoculated against all types of tropical diseases, and warned that we faced lethal threats from snakes, poisonous spiders, and insects that burrow beneath the skin to lay their eggs. Everyone was issued with a set of jungle boots and cotton clothing, which in the armed forces of the 1980s was the best gear going.

Bravo Company flew out to Brunei in May 1987. The draining humidity of the tropics was like nothing I had experienced before. By the time we arrived at Sittang Camp, which sits on the edge of the jungle adjacent to the warm, clear waters of the South China Sea, I was left in no doubt that the next six weeks of jungle training were going to be very tough. More than anything else it was the heat and humidity that were going to kill us – none of the younger guys had ever felt anything like it.

It was key that everyone acclimatized as quickly as possible, and this was supposedly accelerated by punishing early-morning runs along the almost perfectly white, sandy beach just outside the camp. The first run was the toughest, and began at 5 a.m., just as the sun was rising. We ran as a squad along the beach in deep soft sand, stopping occasionally for the stragglers to catch up and being ordered to do push-ups until they did. It was same every morning for the first week, and each session became longer and more demanding.

After an intense session beneath swaying palms trees, we would plunge our sweaty bodies into the warm tropical sea and attempt to cool down before we headed into the jungle for lessons on tactics and survival. Most of us got to grips with life in the dense jungle fairly quickly. It was an intense, unforgiving environment, which you just had to accept and not try to fight against – in that respect it was similar to the Antarctic. No matter how much anyone complained, the bugs were always around, as were the heat and the rain. But the jungle warfare instructors just went with the flow. They somehow seemed resistant to biting ants or the thick clouds of mosquitos that never gave you a minute's respite. Their attitude was impressive, and I tried to follow their lead.

The company was split up into small patrols, and we would be led deeper into the thick and at times impenetrable trees where we'd be taught how to survive and fight in the jungle. We learnt how to set up harbours – temporary bases that we could secure and sleep in at night, in hammocks hitched to the trees. There were lessons in jungle patrolling and ambushes and, as the weeks passed, we found we could move almost silently through the trees. The dense jungle made navigation extremely difficult, and at times our progress seemed to slow to just a few yards an hour – something I would experience in Antarctica many years later, in an entirely different but equally demanding environment.

Halfway through the course we were given a presentation by a US Marine Corps sergeant major, who was a jungle warfare expert, on the latest developments taking place across the pond. The lecture took place in what was called the Jungle Amphitheatre – a natural clearing where students sat on a raised bank where they could watch and listen to one of the instructors giving a lesson or view a tactical demonstration.

Unbeknown to us, the US instructor had rigged up a series of explosives in the treeline about 300 yards away from where the company was seated, known as an A-type ambush. Mortar rounds had been strapped to trees with detonation cord, grenade necklaces had been rigged up and a couple of kilograms of plastic explosives had been placed against laden fuel cans to provide extra dramatic effect.

The instructor stood in front of us all, holding an M16 rifle equipped with a grenade launcher.

'Right, guys, I want to show you this new piece of equipment the US Marine Corps has spent years developing,' he explained to his captivated audience.

'This, gentlemen, is top secret,' he added, showing us what looked like a standard 40 mm grenade, little bigger than a golf ball. 'It is a thermonuclear grenade round. Our scientists back in the States have been working on this since the end of the Vietnam War, when it became apparent that we needed something very special to defeat the enemy in the jungle environment.

'After years of experimentation, and at a cost of several hundred million dollars, we have finally achieved our aim, and I would like to provide you with a demonstration.'

Back in the late 1980s, the equipment available to the British armed forces was at best poor. The US equipment, on the other hand, was what we described as being 'Gucci', and so we didn't doubt his word for a second.

'Now for the demo,' he added, popping the grenade into the launcher.

He fired off the round and we all watched as the grenade arced gracefully into the air and dropped out of sight into the treeline. As it landed, he – or someone in on the gag – detonated the ambush, and the entire jungle across a frontage of around 600 yards exploded into an almighty fireball. The ground shuddered and we could feel the blast wave from where we were sitting. We all sat there completely aghast at this incredible weapon system the US had developed, until we noticed the smile starting to creep across the US Marine's face. 'Hey, guys, got y'all.' Then the penny dropped. One nil to the USMC.

The Royal Marines' role was and remains that the Corps should be capable of fighting in all environments from the Arctic to the desert, and everywhere in between. Jungle warfare was handed to 40 Commando, who were nicknamed the 'sunshine commandos', and we were expected to become masters at it.

Key to surviving in the jungle was excellent personal administration. Small cuts can quickly become infected, feet, armpits and the groin can develop rashes and skin lesions, and the body can at times appear to degrade overnight. The potential for contracting some hideous disease is rife, and the threat from venomous animals, such as the Borneo keeled pit viper, whose bite can kill, is very real. Over the weeks of training, the cotton camouflage clothing and special jungle boots that we had craved for so long began to fall to pieces, as though they were being consumed by the jungle. Trying to keep our feet dry and prevent 'immersion foot' – a debilitating condition caused by marching in permanently soaked boots – was a major battle. We were constantly thirsty, and the acrid smell of weeks-old sweat would cling to our bodies. The heat and humidity tended to stifle the appetite, and everyone began to shed kilograms in weight.

The one benefit of jungle warfare is that nothing happens at night. The sun sets quickly in the tropics and movement through the impenetrable forest is impossible, so several hours of good restorative sleep are usually guaranteed. Night routine began with changing out of our stinking, sweat-soaked clothing into dry kit before climbing into hammocks. As the sun descended beneath the horizon, the jungle came alive with the nocturnal animals that we rarely saw. Large bats would swoop down from the upper canopy in the search for insects. Occasionally large cats and boar would prowl close to the harbour position in search of food, mistaking our body odour for some wild animal.

In the middle of one night, while the company was encamped

in a harbour position deep in the jungle, a series of terrifying shrieks suddenly erupted. Like everyone else, I had been asleep in my hammock when I was woken by the commotion.

'It's eating my brain, it's eating brain,' someone screamed.

My first thought was that someone was having a nightmare, but then I heard movement and saw through the darkness someone running around, falling and stumbling. People started getting up and putting on head torches, quickly realizing that this was something serious and unplanned.

In the middle of the chaos, I spotted a young Marine holding his head and screaming at the top of his voice. It was like a scene from a horror film.

'Help me, help me. It's eating my brain!'

Just as he was about to run off again, two burly NCOs rugby-tackled him to the floor. A company medic appeared and attempted to calm him down.

'My ear, my ear, something's inside my ear,' he shrieked.

It later emerged that a baby hornet had crept into his ear while he slept, and when trapped began to sting his eardrum. The pain must have been appalling.

The medics held him down and tried to flush out the insect using syringes of water; this did eventually work, but only after twenty excruciating minutes. The unfortunate Marine was given a heavy dose of painkillers and extracted back to a Brunei Hospital for further treatment. From that night on I decided that I was going to sleep with balls of flannelette, a form of linen cloth used to clean weapons, in my ears.

The six-week exercise ended with a few days of adventure training, based around Sittang Camp. One of the activities everyone wanted to try out was waterskiing on a small river behind the camp. The company was split into groups, given a quick demo on how to use the boat and off we went.

My group were slotted in for a couple of hours in the afternoon. Each one of us had a few goes up and down the river while the rest relaxed inside the boat or sitting on the nearby jetty. Being one of the youngest, I was naturally last to have a turn. The boat sped up the river, turned around and zoomed back down, bashing into its wake. It was great fun and I was having the time of my life skiing behind, lapping up the sun, and finally believing that some of the stuff described in the recruiting brochures was actually real.

The boat turned and hit its wake again, but this time with a bit more force, knocking one of the guys, Craig, an eighteen-year-old Marine, clean off his seat in the bow. The boat passed over the top of him and as he surfaced he began screaming. As the boat drew to a stop, I came to rest beside him, and to my horror saw the water was rapidly turning red. I immediately realized we had a major problem and that it was likely he'd been hit by the speedboat's spinning prop.

The poor guy was in shock and the look on his face was of pure terror. He knew he was injured, but we had no idea how bad it was until he was lifted out of the water by the rest of the guys in the boat. One of his legs had been severed below the knee and was only attached by a small flap of skin. There were also deep, open wounds in his other leg, which had been torn into by the prop blades. Everyone was in a panic; I was trying to manhandle him up out of the water, while the others were trying to pull him into the boat.

In a matter of seconds, a fun afternoon had turned into a disaster. Back at Sittang Camp, we discovered that there were no medics around, so Craig was placed in the back of an Army four-ton truck and we had to drive along bumpy jungle tracks for an hour to the nearest town where there was a hospital. We stemmed the blood flow with towels and a tourniquet and attempted to offer him words of encouragement, but we knew we were in a race against time.

It was an incredibly traumatic experience and was the first time I had witnessed anybody sustain life-changing injuries. The incident reminded me of my near-miss with the speedboat in Stonehaven Harbour years earlier when I'd tried out scuba-diving. A shudder ran down my spine as, for the first time, I realized how lucky I'd been not to be killed.

A few days later, as we packed up our kit for the return journey home, we were given news of Craig. Although he had been seriously injured and had lost a lot of blood, he had survived. Sadly for the young Marine, though, his leg had been amputated.

We were preparing for an operational tour of South Armagh when Craig reappeared at 40 Commando, on crutches as he prepared for his medical discharge. He later went on to become a successful businessman and a demon downhill adaptive skier.

Fortunately, there was little time to dwell on events. Northern Ireland tours were always preceded by an intense six-week training package down on the south coast of Kent in Lydd and Hythe.

I was happy to be busy again, being put through my paces on the ranges and in various scenarios, and getting up to speed on the latest tactics the IRA were using in their attempt to kill members of the security forces. This was standard practice for troops being sent over to Northern Ireland; we needed an up-to-date understanding of the IRA threat we were going to face. We were flown out to Northern Ireland in February 1988, shortly after my nineteenth birthday, our destination Forkhill, a small village situated less than a mile from the border in a republican stronghold. The threat from roadside improvised explosive devices was so great in the area that the only safe way to travel was either on foot or via helicopter.

The South Armagh IRA had a fearsome reputation. Unlike other IRA brigades across Ulster, it was claimed that the South Armagh IRA had not been breached by the British intelligence services. Whether that is true or not is open to debate, but they were a well-drilled, well-trained organization, capable of launching complex terrorist attacks against the security services.

As we flew into Forkhill on an aging Wessex helicopter, approaching the security forces base, a member of the aircrew turned and shouted: 'There's been a contact. Forkhill has been mortared. There are multiple casualties, we are going in to assist.'

Staring out of the window, my pulse rate quickened as I saw lifeless bodies lying close to the Helicopter Landing Site (HLS) and what seemed to be smoke coming from a building.

One of my mates turned to me ashen-faced and wide-eyed and said: 'Jesus – there are bodies everywhere. We need to get down quickly and help them before another attack comes in.'

The chopper landed and a panicked RAF loadmaster began shouting, 'Off, off, off.'

We poured off the transport as quickly as possible and the Wessex departed seconds later, having barely touched down. As it spiralled away, we ran over to the first set of casualties, some of whom were writhing in obvious agony and were covered in blood-soaked bandages. Then, almost as one, they jumped up and began laughing at us for falling for the wind-up. Talk about black military humour.

Like many other Marines on their first tour of South Armagh, I was initially quite anxious. Our training had taught us to be wary of everything and everyone. Never walk on paths, never use gates, never pick up any discarded items of military equipment, never use culverts – all were at risk of being booby-trapped – and always make yourself a hard target, dashing and zig-zagging out of a military base when going on patrol.

The local people were at best indifferent to our presence; a few were openly hostile and would spit on the ground or call you British scum. Occasionally a car would be stopped at a vehicle checkpoint and the driver would be a known member of the IRA – a 'player', as they were known. More often than not they would be unfailingly polite and cooperative and would sometimes even strike up a conversation about the weather or ask you where you were from in England and whether you were enjoying your time in Ireland – never Northern Ireland. After we'd searched the car and carried out a personnel check, they would be on their way, ready to plan another attack, possibly against us.

Back then the IRA held sway in the border region of South Armagh. Mocked-up road signs stating 'Sniper About' warned us of the ever-present terrorist threat, but most of the time it was a battle against boredom and complacency, both of which could lead to fatal mistakes.

The area 40 Commando operated in was deeply rural and dominated by the mountain of Slieve Gullion, which at 1,880 feet was the highest point in the region. It loomed over a patchwork of fields belonging to small isolated beef or dairy farms, broken up by occasional woodland.

The hardest part of that tour was the weather. It seemed to rain incessantly, and we would patrol for days, soaked through to the skin, crossing mile after mile of sodden, muddy fields, fording rivers and streams, sleeping in woods, waking early before first light with our wet clothing often frozen stiff and waiting for the terrorist attack which never came.

Then, one day, intelligence came through that an IRA Active Service Unit was planning to attack one of the Romeo Towers – fortified observation posts situated along strategic points of high ground along the border. My troop's task was to bolster the security for the towers. To add extra spice to the mission, we were also informed that an operation like this would usually be left to other specialist units, but they were busy elsewhere in the province.

Every night for a week we would smear our faces with camouflage cream, carry out rehearsals, receive a detailed set of orders and deploy onto the ground. The skills we had learnt months before in the Brunei jungle were now being used on the damp, cold mountains of South Armagh.

But each night turned into a no-show and, just before first light, the troops would withdraw and patrol back to Forkhill, exhausted and frustrated. This went on every night for a week without any sight of the IRA – or anyone, in fact.

After a week, the nightly operation was called off. The intelligence was probably a bit sketchy and we quickly came to the conclusion that the specialist units probably looked at it, saw that it was a bit thin and decided to leave that one to us.

The unit returned home in the summer of 1988. My troop had not experienced any direct contacts with the IRA, but we had not lost anyone either. I received the Northern Ireland campaign medal, my first, and with the arrival of more new recruits was no longer regarded as the FNG. I had successfully completed my first Op tour, which is what I joined the Marines to do, and best of all I had managed to save enough money to buy a motorbike.

Within days of being sent home on leave, I had blown virtually my entire savings on a £3,000 second-hand Kawasaki GPZ 600R motorbike – similar to the bike ridden by the actor Tom Cruise in the Hollywood movie *Top Gun*. Owning the bike was like a dream come true, but a week later it was reduced to a heap of worthless scrap and I was about to learn a very tough life lesson.

I had ridden bikes before and had managed to pass the first of a two-part motorbike test, but I hadn't quite got around to taking the second part. This small fact worried me not.

I was returning home to Whaplode from a friend's house in Spalding late one night. It was around 11 p.m. and I was getting more confident on the bike, having ridden it excessively over the previous few days. As I zoomed around one corner, I was aware that I was approaching a level crossing which, in all the years I had lived in the area, I had never seen in operation. But as I emerged from the bend at around 60 mph, the barriers were down and the lights were flashing. In that split second I knew that if I careered into the lowered pole, I would in all probability be decapitated, so I threw the bike onto its side and proceeded to slide underneath the first barrier with the bike in front of me. I then watched as the bike cartwheeled, almost as if in slow motion, and crashed into the far barrier, ripping it completely off its mounting, while I carried on sliding down the road.

I was only wearing a flimsy leather fashion jacket and some tracksuit bottoms, and my knees, hands and back were shredded by the hard, gravelly road. I lay in the dark for around ten minutes, unable to move and praying for the almost crippling pain to subside. After what seemed an eternity, a car came along. Inside were a couple and their daughter who were just returning from holiday. The glare of their headlights must have lit up my prostrate body lying in the road, and they quickly jumped out to help me. Seeing the state I was in, they insisted on taking me to hospital. The little girl who was sitting next to me had probably never seen so much blood. She was crying and asking her mummy if I was going to die.

Although none of the wounds was life-threatening, the damage to my body was extensive and severe and I was kept in hospital overnight. The following morning the police arrived and I feared the worst.

'Mr Rudd,' the traffic officer said to me, 'we understand you have been in an accident and collided with the traffic barrier. We will need to see all of your documents, licence and insurance – it's just a formality.'

My thought process went into overdrive and I knew I was in trouble.

'I don't have them,' I responded quickly, adding unconvincingly, 'they're under the bike seat.'

I can't be sure, but I think even then the policeman suspected that I was lying.

'No matter,' he said. 'We've recovered your bike, so we'll go and have a look.'

A few hours later the same officer returned.

'There seems to be some sort of mistake, Mr Rudd. All we

could find was this provisional licence. And as I'm sure you are aware, you can't drive a bike like the one you were on with a provisional licence – can you explain this?'

I was, as the saying goes, 'bang to rights' and I quickly confessed.

The officer explained that I was going to be charged with driving on an incorrect licence and that I was facing a fine, some points on my licence and an increased insurance premium in the future. I was also told to report to a local police station within a week to face the charges.

A few days later I arrived at Spalding Police Station ready to face the music and thinking that my career in the Marines was also in jeopardy, when by pure luck I spotted my uncle's best friend, who just happened to be working as the duty sergeant.

He recognized my name, confirmed who I was, and let me off with just three points on my licence, a £120 fine and a few choice words of advice. Although obviously disappointed with the loss of my bike, I felt as though the whole sorry saga could have been a lot worse.

The incident with the bike taught me that there was a big difference between being a risk taker and being reckless. My dream of being a Royal Marine commando and leading a life of adventure had almost come to a sticky end, and I realized that I had been very lucky.

As the months passed, I became more and more interested in the role of the Special Forces, particularly the Special Boat Service (SBS). I had joined the Royal Marines to go on operations and to experience combat, and the closest I had come to that was Northern Ireland. The Special Forces seemed to be on operations all the time, around the world, and I wanted a part of the action.

In June 1989, a senior NCO gave a lecture to members of 40 Commando interested in joining the SBS, outlining the role of the SBS and providing details of a recent exercise. He then explained that they were recruiting General Duties Marines – of which I was one – to join as Maritime Counter-Terrorism Assaulters.

Crucially, for me at least, he also said that joining in a supporting role would provide a perfect stepping stone into the world of the Special Forces. I was sold, completely, and the following morning volunteered for the two-week Maritime Counter-Terrorism Assaulters selection course.

The course, which was held in Dorset, was fantastic. There were lessons on specialist weapons that I had never used before. There was an opportunity to conduct assaults dressed in the counter-terrorism black uniform, which I had last seen when the SAS stormed the Iranian Embassy to end the 1980 siege. This was a pretty cool experience for everyone involved, and for the first time I felt that I would be learning some very special military skills.

Most mornings began with physical training, and there was great emphasis placed on fast-roping – sliding down a thick rope in full kit using only your hands, protected by thick leather gloves, to control your speed.

At the end of the two-week course, the names of those who had passed were read out by the chief instructor. Mine was amongst them, and I felt seriously pleased with myself. Best of all, Paul, an acquaintance from 40 Commando, had also passed. We celebrated our success that evening with several beers, and the following morning returned to Taunton with a certain swagger in our step as we arranged our departure to RM Poole.

Life in Poole couldn't have been more different to a regular unit. The ethos was that we would have to be ready for any emergency 24/7, and consequently we trained all the time. Physical fitness was a huge part of the regime and everyone was expected to approach all forms of training with the utmost professionalism.

From the movement I arrived, I was placed on training courses. I earned my Parachute Wings at Brize Norton, which was good fun, and also completed the highly demanding six-week-long Royal Marines Sniper Course. Sniping isn't just about being a good shot. Snipers are trained in camouflage and concealment, judging distance, observation and stalking. Good sniping is all about being patient and professional and having the ability to hit your target with every round.

For the first time in my military career I really felt I was going places. Almost every week I was learning a new skill, and I was still only twenty years old.

Unfortunately, some of the SBS guys treated supporting members with a little contempt, referring to us as 'Dope on a Rope'. The SBS guys would rarely chat or even acknowledge the presence of anyone who wasn't badged SBS. As far as I was concerned, we were all part of the same, well-trained team.

The attitude of some of the SBS aside, life was great. The exercises were extremely well planned and realistic, and most weeks the unit was operating somewhere in the North Sea or English Channel. One exercise nearly ended in disaster for me personally and was a lesson in how events could quickly unravel.

The exercise scenario was that a container ship in the English Channel had been hijacked, and our role was to retake and clear the vessel. The crew were still running the ship for safety reasons but there were military role players on board acting as terrorists.

The plan was to assault at night, but to make matters just that little bit more difficult, the weather was bad and was fast deteriorating. The wind was howling, the rain was torrential and the ship was bobbing around in the sea like a cork in a pond. The team flew in by Chinook and one by one began to fast-rope down onto the ship's deck 90 feet below. Just to add to the realism, everyone was kitted out in full counter-terrorist gear, including a respirator with tinted lenses that, in the pitch-black night, seriously restricted our vision.

Within a couple of minutes, the bulk of our team had landed safely on the ship's deck, leaving me as the last man down. As I fast-roped 90 feet down onto the ship's deck, I didn't realize that the Chinook had moved out of alignment with the actual landing point, and the rope was now running down the outside of a huge funnel that was belching out thick black smoke.

There was no turning back, though. As I slid down at high speed, my legs disappeared inside the funnel and my hands were ripped away from the rope. I grabbed for the side of the funnel, getting a death grip on the lip, and eventually my scrabbling feet found a ledge. I tried to climb out, but the sides of the funnel were too smooth and I couldn't find any higher purchase for my feet. As I looked up, I could see the loadmaster in the door of the chopper, pulling the rope in and watched with a mixture of fear and panic as the helicopter began to peel away.

I began to frantically flash my torch at the loadmaster, who eventually saw that I was hanging on to the inside of the funnel, which I now began to realize was extremely hot. He dropped the rope again and signalled to the pilot to manoeuvre the Chinook over to me. Terrified, I grabbed the rope and after several attempts was lifted out of the funnel and back onto the deck. By the time I got inside the ship and connected with the rest of the team, the vessel had been cleared and secured.

'Where have you been, Lou?' the team leader demanded.

'You won't believe it,' I explained. 'The RAF dropped me into the funnel. I thought I was going to cook. They only came back because I managed to signal the loadie.'

The rest of the team broke into fits of laughter.

'Right,' said the team leader. 'From now on you are going to be known as Sooty.' Fortunately, the name didn't stick.

After hearing about what had happened to me, one of the ship's engineers told me that I was incredibly lucky; had I not caught the rim of the funnel, I would have fallen onto a grate about 20 feet below, where I would have been effectively slow-cooked by the heat from the engine.

Despite this mishap, as the months passed, I was increasingly convinced that I wanted to become a member of the Special Forces. I had joined the Royal Marines to take part in operations, and I quickly came to the conclusion that I had far more chance of that as a member of the SBS rather than serving in a commando unit.

In 1991, the First Gulf War kicked off and I watched with envy as the majority of the SBS left Poole for the war. It was clear that our UK counter-terrorism role meant we wouldn't be going anywhere.

But the First Gulf War was not designed for SBS operations – and in many ways it was a non-event. News began to filter back that while SAS Sabre Squadrons were marauding around the desert attacking Saddam Hussein's Scud sites, the SBS did very little.

Paul, who I'd served with in 40 Commando, became my best mate in the unit. We both wanted to join the Special Forces; after the Gulf War we both decided that we would apply to join the SAS, where we felt we'd be more likely to see action than in regular units. It was a decision that we knew would meet with some resistance, given that we were both Royal Marines and part of the Royal Navy and the SAS was part of the Army.

'The SAS? Why do you want to be a pongo?' was the gruff response from the Scottish SBS sergeant major when Paul and I put in our requests to attend SAS selection. 'If you want to join the Special Forces, you should be applying for the Swimmer Canoeist 3 Course'; this was the name of SBS selection at the time.

There was no way that either of us was going to give our real reasons, so we both sort of invented medical problems.

'I can't clear my ears under water, sir,' I said with convincing confidence. 'I joined the local sub-aqua club and it's a problem that can't be resolved.' My excuse wasn't completely invented. It was a fact that I struggled to clear my ears under water, but I wasn't a member of a local club.

'And I'm partially colour blind – I can't distinguish between red and green,' Paul told him, 'so I wouldn't be able to distinguish between different navigation buoys.'

But our excuses carried little weight, and we were both chinned-off by the sergeant major. 'I'm not letting you go anywhere for a year,' he told us. 'There are only a few slots a year for Royal Marines on SAS selection and those have already been filled. Besides, we are too short of men and we spent a lot of time training you guys – we want to get our money's worth. Come back in twelve months. Close the door on your way out.'

Paul and I were completely deflated. We both felt that our careers were being stifled by some form of petty inter-services Special Forces rivalry, but neither of us was the type to kick off or be resentful. We carried on working hard on training exercises and keeping our heads down.

In the year that followed I met my future wife, Lucy. She was a local girl from Wareham, working at Barclays in Poole, and we met on a night out. Lucy and I hit it off immediately, despite me getting a bit drunk on only our second date and leaving her on the beach holding my clothes while I insisted on jumping off Bournemouth Pier for a midnight swim.

We dated for a year and then decided to buy a place together. I was living in a small, dingy room in the camp, and Lucy wanted to move out of her parents' home, so it made sense to buy a place together near to RM Poole. Paul moved in with us and rented a room, and hardly a day or evening would pass without some sort of conversation about SAS selection. It became something of an obsession for both of us.

A year later we both submitted our request again, and this time we were called forward for an interview with our squadron commander. 'What have you got to offer the SAS as a twenty-one-year-old Marine with limited operational experience?' was his first question as I walked into his office.

'I'm para- and sniper-trained, sir,' I responded rather meekly. 'And, sir, I was told to wait a year and that is what I have done.'

'If you fail in the first week, you will be an embarrassment to the entire Royal Marines. You are taking a big risk. It will have serious implications for your future, so you need to think very carefully about this.'

'I appreciate that, sir, and I have given it a lot of thought. I have done exactly what I've been asked to do and I can assure you that I will not embarrass the Corps.'

'In that case – good luck', he added. 'You have a lot of training to do and you will have to fit that around a very heavy work schedule.'

I saluted, did an about-turn and could barely contain my delight. Paul followed me in and was also granted permission. All we had to do now was pass the world's most demanding Special Forces selection course and we'd be members of the SAS.

5

PILGRIMS' PROGRESS

We are the Pilgrims, master; we shall go
Always a little further; it may be
Beyond that last blue mountain barred with snow
Across that angry or that glimmering sea.

James Elroy Flecker

'Most of you will fail', the SAS's Training Squadron chief instructor announced without a flicker of emotion as he addressed the 120 volunteers for SAS selection. 'Some of you will fail tomorrow. Some of you will get injured and some of you will just give up.'

I swallowed hard and shifted uncomfortably in my seat.

'There is no point at this stage in getting to know any of you by name, because by the end of the course we expect only around ten to fifteen of you to pass', the chief instructor added. 'Don't take it personally, and maintain a sense of humour and humility.'

Up until that moment, I believed I was ready to undertake the British Army's most gruelling form of physical and mental selection. Now I had been stung by self-doubt and wondered whether I had done enough. I stole a glance around the room. Everyone looked apprehensive. There were no smiles, only stern faces. My heart began to race as I contemplated the reality of failure. The thought of returning to the Marines having failed to make the grade left me feeling physically sick.

I had seen what happened with other guys who had returned early from SBS selection. The general view, often unfairly, was that they had 'jacked' – given up. The ribbing was unrelenting and could last for weeks. Some never quite managed to escape the shadow of failure, and it could dog them for the rest of their careers.

I too had been repeatedly told that I was too young and too inexperienced, but I thrived on that negativity. I always have done. I've never listened to anyone who has tried to convince me that a physical challenge was beyond my capability. I wasn't super-fit but I knew I had an inner strength, a self-belief that I wouldn't be beaten by a physical challenge.

It was January 1992 and I was just twenty-two years old, easily the youngest man in the room. I was surrounded by battle-hardened Army corporals and sergeants, veterans of the Falklands and Northern Ireland, all infinitely more experienced than me. My ability to fast-rope from a helicopter onto an oil rig or the deck of a rolling ship in rough seas counted for little now.

Passing SAS selection had been the complete focus of my life for more than two years. Now I was struck by the daunting realization that the odds of passing were – at best – very slim. I said it over and over to myself: ten passes, 110 failures.

Six months earlier, Paul and I had begun training for SAS selection. If you wanted any chance of passing the arduous month-long hill phase of SAS selection, you had to invest a lot of time in the Brecon Beacons national park, the Black Mountains, and the remote Elan Valley, a place of haunting beauty where it only seems to rain. The area has been the SAS's testing ground for more than sixty years. It is a remote mountainous region containing some of the toughest climbs in Wales – perfect for separating the wheat from the chaff.

We were two impoverished Marines – a rank equivalent to an Army private – and rather than spend a small fortune on local B&Bs, we spent £300 on a small, clapped-out, mouldy caravan whose better days had long-since passed. There was no heater or cooker and the toilet was a bucket in the corner.

Every Friday afternoon, after a week of hard training with the troop, we hitched the caravan to Paul's car and towed it from our base at RM Poole up to South Wales and parked it on a quiet country lane or lay-by where we hoped it wouldn't be removed by the police or an irate farmer. It wasn't comfortable but it was cheap, and it would serve as our training base.

At the end of each day of yomping up through the steep, treeless valleys of the Beacons, wading through foaming streams and rivers, we would collapse exhausted in a muddy, sweaty heap. We cleaned ourselves with a bucket of hot water and repaired our blistered feet with plasters and bandages purloined from the medical centre back at Poole. Dinner was a takeaway or, if money was tight, a twenty-four-hour ration pack, and we were often asleep before 9 p.m. The same punishing schedule would begin again early on Sunday and would end with a 140-mile drive back down to Poole to begin the working week.

By the time the selection process began, we were confident about navigating our way around the Beacons during night and day and in all weather conditions – sun, hail, fog and rain. In winter the hills had become treacherous. The ground on the higher peaks would freeze at night and was as slippery as polished ice the following morning. Progress was often made on the steeper slopes by crawling on all fours. By the time selection had arrived, we had carried loads of up to 30 kilograms over many miles. We were fit, determined and brimming with confidence.

Two weeks before selection was due to start, I received a copy of the deliberately vague joining instructions. The single sheet of A4 paper shed little light on the trials ahead, apart from a warning that candidates would have to pass the Army's Combat Fitness Test – an eight-mile march to be completed within two hours, carrying 15 kilograms of equipment together with a rifle and helmet – which for anyone contemplating SAS selection should have been a walk in the park.

There was also a short kit list and directions to Sennybridge Camp in Wales, an isolated training camp composed of a collection of pre-war buildings not designed for comfort.

On our first day, in early January 1992, we gathered in a large rectangular room lit by buzzing fluorescent lights, which seemed to serve as a classroom. At one end was a blackboard and a small stage. None of the windows was open and – either by fault or design – it was uncomfortably warm. The atmosphere inside was fetid and thick with the smell of stale breath and sweat. It was the smell of fear.

The chief instructor scanned the room: 'We are looking for potential.' He wore the sand-coloured SAS beret and the Regiment's sky blue stable belt and had the composure of a man who had seen much and feared little.

'No one is a ready-made member of the SAS, or the finished article,' he added. 'We expect you to be professional in everything you do. At this stage it's about doing the basic soldiering skills to a high standard.'

His sentences were short and clipped and I hung on every word.

SAS selection was the first time in my military career that I had ventured out from what to me was the closed world of the Royal Marines. Sitting in that classroom were soldiers from across the Army. Welsh, Irish and Scots Guards. Sergeants and corporals from regiments I had never previously encountered, like the Black Watch and the Green Howards.

And then there were the Paras, the Marines' bitter rivals. The Paras sat in a group separated from everyone else. Most had thick, bushy moustaches, their famous red berets pulled down low over their eyes. They were a group apart, and had a fearsome reputation for holding all other Army units in complete contempt. The Paras believed in themselves and each other. Everyone else was a 'craphat' – the Marines included. The Paras made up almost 40 per cent of the SAS's manpower, and the attitude of some was that they had a divine right to be there. As for Marines, we were something of a novelty.

After the chief instructor's brief, we were shown our sleeping quarters – long, dormitory-like rooms lit by more buzzing fluorescent lights. Everyone was allocated a bunk-bed space and a rusted steel locker. The rooms were overcrowded but the SAS instructors – called directing staff or DS but known to us as 'Staff'– cheerily assured us that there would soon be plenty of room.

Sleep was almost impossible that first night. The chief instructor's words were bouncing around my head, and the kit fiddlers – nervous soldiers constantly adjusting their equipment in preparation for the following morning's first test – kept me awake until the early hours.

'Guys,' I pleaded, 'your kit's fine. It's just a CFT, you need to get some sleep.'

But my advice was ignored. I was a twenty-two-year-old Marine and some of them were sergeants in infantry regiments with many more years of service. They weren't going to listen to me.

By 8 a.m. the following morning, all 120 of us were on the start line in the hills above Sennybridge waiting for the 'go.'

'Anyone who fails will be immediately returned to their unit,' the DS in charge of the test reminded us.

Astonishingly, there were a couple of failures. How anyone could turn up for SAS selection and not be fit enough to pass a CFT was beyond me.

The following day was a much tougher test – the notorious Fan Dance. It didn't matter how fit you were, the Fan Dance was designed to hurt.

The course required the towering 2,907-foot peak of Pen y Fan – the highest and most dramatic mountain in the Brecon

Beacons – to be climbed twice. Up and over, a quick turnaround and up and over again. Every man carried 15 kilograms, bergen and weapon. The pace was blistering and was set by a super-fit member of the DS who clearly could not feel pain. Almost immediately our ranks began to thin. The first leg felt like a near-vertical ascent of the mountain, and more like a charge than a march. By halfway, people were dropping like flies, vomiting and feigning injury.

The DS set the pace and we were expected to stay with him. After one half had been completed, we turned around and followed our steps back over the mountain down to the start point. Anyone who did not complete the course in four hours thirty minutes was RTU'd – returned to unit – no excuses.

The Fan Dance sliced through the course like the Grim Reaper's scythe, and by the end of the second day more than a third of those who had arrived on Sunday evening had been lost through injury or what was called a voluntary withdrawal (VW). My hut was now half full and the kit fiddlers had gone.

After the Fan Dance, life calmed down. The daily marches, although testing, no longer verged on the impossible. The challenge now was to stay injury-free. Every day for the next three weeks a dwindling number of SAS hopefuls took to the hills. The distances increased, as did the weight we carried on our backs. We marched in small groups, then in pairs and finally on our own, navigating in all weather conditions, by day and night. I felt I was doing well and putting in a solid performance. The time invested in training was paying off: as the saying goes, 'failing to prepare is preparing to fail'.

The DS were constantly watching, taking notes, forming opinions. No words of encouragement or criticism were offered. Self-discipline was the order of the day.

My strategy was to get through each day, one hour at a time. If I was still on selection by the end of the day, I was doing well. There was absolutely no point in looking beyond the next twenty-four hours. Most evenings were spent trying to consume as many calories as possible by gorging on pizzas, kebabs and fish and chips, and back in those days we were convinced that a pint of Guinness in the evening would add some much-needed iron to our bodies. But the dropout rate continued unabated.

Guys who you'd thought were doing well were suddenly disappearing, their faces and names quickly forgotten. By the start of Test Week, the 120 starters had been whittled down to about fifty. Test Week was the conclusion of 'The Hills' section of the training; if I could get through that, I would be on to the next stage.

But my morale had been dented by a knee injury I was secretly harbouring. A slip on a greasy descent down a gulley in the Black Mountains had gone from a minor injury to a barely disguised limp. If I went to the medical centre and asked them to take a look at the injury, I risked being RTU'd, so I'd kept the pain to myself and dosed up on Brufen.

Every evening Paul and I would chat about the day's events and what had gone well and what hadn't. It was great to have a good friend to confide in at all times, especially on the days when your morale might be taking a bit of a dip.

Test Week consisted of five separate marches, and each called for a different set of skills. Each march had a cut-off time and there was no allowance made for weather or injury. Snow, fog, rain – it didn't matter. Beat the clock and you passed, a second over and you failed.

I saw my preparation – the months I had spent driving up to Wales every weekend – as putting money in the bank, a deposit of fitness I could draw on through The Hills phase of selection. But by the time Test Week arrived, my account felt well and truly empty. The demands of selection are such that the body doesn't have time to recover. My joints ached, I had friction burns on my back and shoulders, and my feet were in tatters. The most painful part of every day was peeling away the blood-soaked socks from the flapping skin of my shredded feet.

The weather had also taken a turn for the worst. Snow and ice covered the hills, making walking a treacherous process. A simple slip, a sprained ankle, a twisted knee, and my SAS selection would be over.

The final hurdle of Test Week was the infamous 'Endurance' march – a 40-mile solo trek across the Beacons carrying around 25 kilograms of equipment, plus food, water and a rifle, with a time limit of around twenty hours, depending on conditions. The route wound its way round and over the Beacons from one end to the other, and began at 2 a.m. in deep, soft snow on a dark winter's morning. It was the ultimate test of navigation, self-discipline and sheer grit and determination. After the Marines' 'Endurance' course, this was the second challenge I had encountered featuring this name, and little did I realize at the time how significant this word would become in my life.

Our instructions were simple – finish the march within the allocated time, passing through all the manned checkpoints. We were lined up one behind the other; an orange safety panel tied to the bergen of the man in front was the only visible feature on that black, moonless night. We set off at four-minute intervals. There were no words of encouragement, no, 'You're nearly there, just one last hurdle' – that was not the SAS way. Just a simple, 'Off you go.'

I drove forward, one step then another, climbing higher and higher into the darkness, my legs pumping like pistons. The bergen straps cut into my shoulders, already raw from three weeks of punishment, but I was almost immune to the pain.

To pass Endurance there was an average speed I needed to achieve, dictated by the weather conditions, to make the cut-off time. It was almost impossible to hit this speed going uphill, but lost time could be regained by running downhill and on the flat sections. The plan began to fail after a few hours when the pain in my knee became unbearable. Any attempt to run downhill was beyond agony. For a while I managed to ward off the pain by taking copious amounts of painkillers and by concentrating on navigating and pacing.

When my concentration flagged, I focused on why I was there and the benefits of success. One step at a time, keep moving forward, stay on course, get to the next checkpoint – it became my mantra over the next twenty-odd hours.

Twenty-six years later I would deploy the same technique to drive me towards the South Pole and beyond.

As the sun rose, I could see a string of orange luminous marker panels stretching out in single file across the rugged landscape ahead of me, and although the golden rule was never to follow the man in front, I took some comfort from knowing that others were around.

By early afternoon I had reached Cray Reservoir – the halfway point. I refilled my water bottles, demolished a Mars Bar and set off for the second half of the march.

As night fell my energy levels plunged – I was running on empty. I fell several times and began to hallucinate. My body was crying out for sleep, and for the first time I really feared that I might fail the final hurdle. The thought of failure filled me with fury. I had come so far and now – through a lack of determination – was I really going to allow my dream to slip away? I focused on the pain in my knee and pushed on.

It was dark once more when I reached the final checkpoint. I had beaten the clock. Under twenty hours and the pain had disappeared. I was buzzing. A few minutes later I saw Paul emerge from the darkness. He had made it too.

I had passed The Hills phase, and there were just thirty-three of us left. We were treated to a weekend off; on the following Sunday we reported to Stirling Lines, the home of the SAS in Hereford. Although I had a long way to go before completing

SAS selection, the very fact that I was now living within the SAS world felt good.

The thirty-three of us who had passed the first phase had a separate barrack block from the rest of the SAS soldiers and our own seating area in the cookhouse. Everyone knew who we were – they could tell by the fact that we wore our own uniforms – but the atmosphere was different to Poole. It was more relaxed and informal. And although the badged SAS guys were obviously told not to have any contact with us, they acknowledged our presence with the occasional hello or a nod and a wink as if to say, 'We've all been where you have, mate – so good luck.'

The next phase of SAS selection was what can be euphemistically be described as 'advanced military skills in a challenging environment' – put simply, pure soldiering. For reasons of operational security, I am not allowed to reveal where this took place, but I can say that it was and still is the most demanding of all operational environments. The course was designed to push candidates to their mental and physical limits in the most arduous combat conditions imaginable and it succeeded in achieving this on every level. The environment was tough, unforgiving and ruthlessly exploited any and all weaknesses. Absolute professionalism was required just to stay healthy and able to function, let alone train to fight against a well-equipped enemy.

Alongside the rigours of the environment, we had to very quickly learn and adapt to new tactics on a daily, sometimes hourly basis and always under the ever-watchful eye of the SAS DS. The only respite came at night when, under the cloak of total darkness, we had time to rest and relax in the comforting knowledge that at some point in the future the phase would end. When it did, the number of those who remained on selection had been reduced to just eleven. Now all we had to do was pass a two-week package of combat survival and evasion. It is safe to say that by that stage of the course, those of us left on selection knew we were within touching distance of joining the Regiment, and we thought the next two weeks would be a bit of a breeze. On that score we were wrong.

The first few days of combat survival took place in woods in Herefordshire and, by comparison to the other phases of selection, were interesting and fun. Instructors, some civilian, taught us how to build shelters, make fires, and track, catch and prepare game. There were also lessons on how to survive off berries and fungi. There was a lot of banter and the DS even smiled occasionally.

During this stage, all of the remaining candidates were living in the woods, in shelters that we had built following lessons from survival instructors. In place of rations we were given a rabbit and a chicken to gut, prepare and cook. I soon realized that the lack of sleep and food was meant to soften us up for the evasion phase, which was as challenging as you might expect. And that's all I can tell you!

At the end of the two-week training package, the ten of us who were left gathered together in a classroom and waited for the arrival of the chief instructor. After ten minutes or so he emerged with a box of the famous sand-coloured berets and walked around the room handing them out.

'Congratulations, Rudd,' he said shaking my hand. 'Welcome to the Regiment.'

It was possibly the lowest-key military ceremony I have ever taken part in, but for me it was also the most important.

Humour and humility were two of the tenets of the SAS ethos as laid down by its founding member Lieutenant Colonel David Stirling during the Second World War. Along with the unrelenting pursuit of excellence and self-discipline, they would become the handrails that would guide me throughout the remainder of my military career and indeed on expedition.

Out of the original 120 who began selection six months earlier, just ten of us had passed and two of those were Royal

Marines – me and Paul. That evening, I travelled back to Poole to break the news to Lucy that I would now be based in Hereford. We agreed she would stay in Poole initially until we were married the following year and could move into Army quarters. I also had to go through the technical process of leaving the Royal Marines, part of the Navy, and joining the Army. On the morning of Monday 15 June 1992, I became a member of 22 SAS.

I was now an SAS trooper, but I was just at the start of my SF career, and still had a mountain of stuff to learn. Although I had passed the toughest military selection course in the world, in the eyes of my peers in 22 SAS I still had a lot to prove. It was a case of keep your head down and your mouth shut, work hard and learn. It went without saying that you were expected to give 100 per cent effort in everything and to be utterly professional at all times – the unrelenting pursuit of excellence.

In the years that followed, I travelled the world, serving on operations and training in some fairly exotic places such as Ecuador, Norway, Pakistan, Zimbabwe and Mauritania. I was on the ground in Afghanistan in 2009, and I learnt how to cope with success and failure and the loss of friends both killed in action and wounded. There was little time to mourn. Friends would pay their respects at the funerals held at Stirling Lines in Hereford, and we would move on and prepare for the next mission. The workload was frenetic, and one operation was quickly followed by another, or by an equally intense and exhausting period of training. In the years that followed, I progressed up through the ranks and became qualified in demolitions, sniping, reconnaissance, surveillance and operating in all environments around the globe, including desert, jungle, Arctic, bush and urban.

I also found that I had a natural flair for leadership. I wasn't bossy or autocratic, I simply enjoyed passing on my experience and helping others to get the best out of themselves. I knew when to encourage and when to cajole, when to ask and when to demand and, when the situation required it, to simply say 'follow me'. But the biggest lesson I learnt from twenty-five years in the SAS was that almost anything was achievable with good leadership, extensive planning and total commitment to the mission.

6

LEADING BY EXAMPLE

Onwards.

Henry Worsley

It was a Tuesday afternoon in June 2009 when an email dropped into my inbox with the heading: 'Centenary Antarctic Expedition: Places Available'. My heart almost missed a beat as I opened the email and read that Lieutenant Colonel Henry Worsley MBE – whom I vaguely knew – was planning to undertake an expedition to the South Pole in 2011. The details were deliberately brief and the short note stated that anyone interested should attend a brief at 4 p.m. on Thursday afternoon in the Training Wing cinema in Stirling Lines.

Henry had something of a reputation within the Regiment as an old-school adventurer. He was charismatic, single-minded, and a man obsessed with the early polar explorers, particularly Shackleton. He possessed the relaxed, easy charm that is typical of many SAS officers I had worked with over the years, and his enthusiasm for all things polar was infectious. He was a man of slight physical presence, unlike many in the Special Forces world, but he was clearly fit, and his piercing blue eyes seemed to reflect an inner strength and self-belief.

He was confident without being overbearing, and I liked him straight away. I didn't know it at the time, but over the next few years he would become a great friend, mentor, and one of the most influential people in my life. I owe a huge amount to him.

Now Henry stood in the centre of the room and announced his plan: 'Gentlemen, in two years' time, it will be one hundred years since Scott and Amundsen raced to the South Pole. I intend to run an expedition called the Scott–Amundsen Centenary Race, and I am looking for volunteers to take part in what was described in 1911 as the "Greatest Race on Earth".'

As he spoke, I was transported back to my childhood, when I stood outside the headmaster's office waiting to be caned and first read about the exploits of one of Britain's greatest heroes, Captain Robert Falcon Scott.

Henry scanned the room as if attempting to get the measure of us all. Were we there out of genuine interest? Did we have a passion for adventure? You could say that Henry had the Antarctic in his DNA, his fascination with the continent sparked by the fact he was distantly related to Frank Worsley (whose father was also named Henry), the New Zealand captain of the *Endurance*, the ship used by Sir Ernest Shackleton's Imperial Transantarctic Expedition of 1914–16 – often described as the last major expedition of the Heroic Age of Antarctic Exploration. Endurance was starting to feature in my life once again.

The concept of the expedition, Henry explained, was to effectively re-enact the 1911 race to the South Pole, with two teams of three men starting from exactly the same locations as the Scott and Amundsen expeditions.

Henry provided a short but compelling history of how, in January 1911, the Scott and Amundsen expeditions arrived in Antarctica and established base camps – Scott's was located at Ross Island, while Amundsen based his camp 400 miles away at the Bay of Whales, on the edge of the Ross Ice Shelf, or the Great Ice Barrier as it was known back then. The bay was named by Shackleton in 1908 during the Nimrod Expedition, because of the number of whales spotted in the area.

Over the next ten months, the two teams prepared for the epic challenge ahead. Fuel and food depots were laid at a series of strategic locations along the routes planned by both men.

Amundsen was the first to set off, leaving Framheim (the name of their base, meaning 'the home of *Fram*', Amundsen's ship) at the Bay of Whales on 19 October 1911. He reached the South Pole on 14 December and arrived back at his base camp just six weeks later on 25 January 1912. Scott, on the other hand, was beset with delays almost from the start, leaving on 1 November and arriving at the South Pole on 17 January, some thirty-three days after Amundsen.

After discovering that Amundsen's team had beaten him to the Pole, Scott wrote in his personal diary:

> The POLE. Yes, but under very different circumstances from those expected. Great God! This is an awful place and terrible enough for us to have laboured to it without the reward of priority.

The return journey must have been truly terrible. Scott's entire team was suffering from slow starvation, hypothermia and various vitamin deficiencies. Frostbite was also beginning to take hold, and two of his team perished en route. First to die was Petty Officer Edgar Evans, followed a month later on 17 March 1912 by Captain Lawrence 'Titus' Oates, who, crippled with frostbite, effectively sacrificed his life in the hope that the rest of the team might make better progress without him.

Captain Oates's selflessness was recorded in Scott's diary:

> 'I am just going outside and may be some time . . .' We knew that Oates was walking to his death . . . it was the act of a brave man and an English gentleman.

The rest of Scott's team, starving and frozen, were to die a few days later, trapped inside their tent by a ferocious polar storm.

Food and fuel that would have kept them alive was at One Ton Depot, just 11 miles away.

In Henry's retelling of the Scott–Amundsen race, he also explained how little had changed in the last hundred years and that the hazards faced by both teams still remained as great as they did back then.

Like everyone who attended the lecture that afternoon, Henry had proved himself on numerous military operations and had been decorated for his service in Afghanistan. But travelling and surviving in the Antarctic, he explained, would require something different.

'It is vital everyone understands that this is not a jolly,' Henry stated, again scanning the room. 'It will be the single most challenging experience of your lives. It will make SAS selection look like a walk in the park. The challenge is also mental, not just physical. It will test you in every way possible. Men – strong, capable men – have been driven close to madness by the extremes we will all face. Antarctica has a knack for exposing the naked soul of a man.

'We will be hauling pulks weighing approximately 120 kilograms for ten hours a day for up to seventy-five days across 900 miles of the most hostile terrain on earth. Unlike the Scott and Amundsen teams, who utilized dogs, ponies and motorized sledges, and laid fuel and food depots along the route, this expedition will be completely unsupported and unassisted from day one until the finish. The temperature can fall to -40°C and we will face winds of up to fifty miles an hour. When we hit a whiteout, as we will, there will be nothing to see for hours, sometimes days, affecting your mind and morale. You will have no idea of where you are or the progress you have made. You may suffer from motion sickness as you try and ski across the ice.

'This expedition will require every ounce of courage and inner strength you possess, and even that may not be enough. We will trace the almost identical routes taken by Scott and Amundsen one hundred years before us.

'The dangers are varied and considerable and are the same as those faced by Scott and Amundsen. Apart from frostbite, snow blindness and malnutrition, each team will have to cross areas of crevassing, some big enough to swallow entire houses.

'We will man-haul our pulks up and over the 12,000-foot Transantarctic Mountains. This has never been done before on the Amundsen route and there are those who believe it cannot be done – gentleman, I am not one of those. Make no mistake, this expedition will be the ultimate test of physical and mental endurance.

'I had hoped that the Norwegians would be interested in re-running the race, but they have declined. As far as I am concerned, the centenary of the Scott and Amundsen expeditions should be marked by another expedition. It will take a great deal of commitment from anyone who volunteers, so you should make sure you get the appropriate clearance from your chain of command before committing. But for those of you who are selected, I can promise you the experience of a lifetime.'

The worse he made it sound, the more I wanted to be part of it.

For the past seventeen years, since first becoming a member of the SAS in June 1992, my life had been an operational roller-coaster. I had been deployed all over the globe, before the world changed with the four al-Qaeda attacks on US soil on 11 September 2001 and the SAS's workload went into overdrive.

As the war against extremism took hold, like many of my colleagues I served in Afghanistan, then Iraq and Afghanistan again. In between that period, I had also managed to start a family. I would also like to say I had raised three children, but that job had largely been left to my wife Lucy, while I took part in a seemingly endless list of overseas operations.

In those seventeen years I had risen up through the ranks from trooper to warrant officer, and had accumulated a vast amount of operational experience. But most of all I felt I had learnt how to lead men in extremely pressurized situations. No matter how dangerous the situation, and this may come down to SAS training, I have never panicked. I have always been able to think clearly, observe, and make the right call. No doubt there has been a lot of luck involved, but some good judgement too.

The idea of visiting Antarctica back then was so far off the agenda that it didn't even rank as a pipe dream. But on that day in June 2009, an opportunity had been presented and I was excited beyond belief.

Henry explained that he would lead one team and that Mark, a highly experienced SAS sergeant major who also had completed a solo journey in the Antarctic, would lead the other. Henry had completed Scott's route to the Pole on a previous expedition so was keen to compare it to Amundsen's route this time round.

Simple maths (the only kind I was good at) revealed that there were just four places up for grabs. Everyone who was interested in the trip and could commit to a two-year training programme was asked to complete a questionnaire detailing their physical health, relevant skills and experience and, crucially, why we wanted to be part of the expedition. My relevant experience was a bit thin. I didn't think parachutist and sniper was what they were really after, and although I had some first-aid knowledge, there were several guys in the audience who were far more medically qualified and had completed specialist medical courses including stints in hospital casualty departments.

It was two long weeks before the next meeting. The numbers present had reduced to just twelve after other members of the Regiment realized the expedition was too much of a commitment.

Henry thanked us all for returning then added: 'We've had a good look at the questionnaires that everyone completed, and we had thought about running some sort of selection process. But you are all members of the SAS, and so we came to the conclusion that would be pointless. The problem for us is that you are all suitable and capable candidates. So, we've decided the only fair way to proceed is to draw names out of a hat – or, in this case, a bag.'

As Henry was talking, Mark wrote down all the names of those present and placed them in a black cloth bag. It was classic SAS organization – very informal and mega-relaxed. The first four names were pulled out and mine wasn't amongst them. I felt incredibly deflated as I saw the look of delight on the faces of those joining the expedition, and was just about to get up and leave when Henry said: 'I suppose we'd better have a couple of reserves just in case.'

Before he pulled out the names, he said that the reserves would have to attend all the training and would also be expected to invest up to £2,000 on skis, boots and clothing in the knowledge that they might not come on the expedition. A few of the guys rolled their eyes in disappointment – it was a big ask.

'If you can't commit to that, and I can fully understand if that's the case, then you'd better leave now.'

The room remained still and the process of choosing the reserves began. I was secretly praying that my name would be called, but also gripped with real fear that my chance was slipping away.

The name for the first reserve was called out – not me. Henry put his hand into the bag for the final time and pulled out a folded piece of paper.

'The second reserve is . . .' Henry seemed to pause for dramatic effect, then said: 'Lou Rudd.' I was elated but tried to keep my emotions under control.

Eventually the meeting drew to a close. I sidled over to Henry as the room emptied.

'Hi Henry, I'm Louis Rudd. Great presentation. I've been fascinated with the Antarctic since I was a kid and it's been an ambition of mine to visit the continent for years. I know I'm only a reserve at this point, but I really want to get involved as much as possible, so if you need any help then please let me know.'

Henry smiled and seemed to be genuinely impressed with my interest in polar history. I certainly felt he'd found a kindred spirit.

He said: 'There are almost two years before the expedition is due to start. I'm sure in some way, by hook or by crook, that if you are still keen you'll be able to come along. Just stick with it, attend all the training. There is a long way to go and I'm pretty sure the names will change, they always do. If I can raise an extra £50,000, then you can come along – it's all about the money. It doesn't matter to me whether it's three or four in a team.'

Later that evening, I rushed home and broke the news to Lucy and my three teenage children. They were all excited for me but also saddened. I had a lengthy operational tour already scheduled, and so would be home for just five months in the whole of 2011. It was impossible to sleep that night and I gave up at around 3 a.m. and went downstairs to the sitting room to read a few pages of *The Worst Journey in the World*, Apsley Cherry-Garrard's frank account of Scott's expedition, while Bailey, my beloved black Labrador, slept at my feet.

Over the next two years there were various training weekends in Hereford, during which we practised crevasse rescue drills on the climbing wall and tried to emulate the feeling of pulling a pulk by dragging one or two 20-kilogram Land Rover tyres across the countryside, using the same harness that would be used on the expedition.

On one particular training session, one of the team joked that the tyre hauling was getting a bit boring.

Henry was not impressed: 'There is no better form of training apart from the real thing. Coping with monotony is part of the challenge. If you can't cope with pulling a tyre around a field for a couple of hours, how do you think you'll cope with dragging a pulk across the ice for weeks on end?'

As well as the training, there were also presentations on the equipment that would be used and the type of clothing needed to battle temperatures of -40°C.

After six months, one of the guys chosen to be in the team was injured in a bar fight while on a squadron trip to Canada. He didn't start the fight and just happened to be in the wrong place at the wrong time. He was glassed in the face, damaging his eye, and although the wound healed, he was advised against going down to the Antarctic, which meant that the first reserve was used and I became the only remaining reserve.

But I suppose my luck changed when a member of Henry's team was ordered to withdraw by his commanding officer after it emerged that he had developed a very severe case of Obsessive-Compulsive Disorder and it was highly unlikely that he would be able to cope with living in a tent for two months with two other people.

While the training was progressing, Henry was working desperately hard to try and raise funds for the expedition. The estimated cost was around £400,000, but Henry was a master at raising finance. He managed to get BP on board, who provided over £150,000. He also got Cadbury interested on the basis that they supplied the drinking chocolate on Scott's original expedition. JCB also provided cash, as did the makers of Chivas Regal whisky. I helped as much as I could and began to get a handle on how complex the planning had become. As well as trying to raise finance, Henry was also negotiating with the Army Adventure Training Group, who would ultimately have the final say on whether the expedition was given the green light.

On top of all the fundraising, Henry was also determined to use the publicity he hoped the expedition would attract to raise over £100,000 for the Royal British Legion's Personnel Recovery Centres. It was a mammoth undertaking, which he had to fit in around a demanding day job at the Permanent Joint Headquarters at Northwood in Middlesex.

While Henry raised the cash, Mark began dealing with the logistics and air freighting, as well as applying for the various permits we would need to have before we got anywhere near Antarctica.

Antarctic expeditions can cost hundreds of thousands of pounds to mount. Those taking part need to be either very rich or have some serious financial backing. The primary company providing logistic support in the region is called ALE – Antarctic Logistics and Expeditions, a very professional American-based organization. Every season, ALE charters a huge Russian Ilyushin transport aircraft and crew for three months. The aircraft, which flies into Antarctica from Kazakhstan, provides a vital airbridge to Chile. As well as the airbridge, ALE establish a tented camp at Union Glacier, on the edge of the Ronne Ice Shelf in Western Antarctica, manned by around forty staff who remain at the base for the entire three-month summer season, which runs from November to January. A fleet of three smaller ski-equipped aircraft are also chartered to ferry passengers deeper into Antarctica.

The ALE operation is a mammoth undertaking and a logistical nightmare, and one of the reasons why Antarctic expeditions are so expensive. Every single item they need has to be airfreighted in at vast expense. The cost of a 45-gallon drum of aviation fuel is a good example of how inflation kicks in. The oil might cost 100 USD in the United States. By the time the same drum has arrived in Chile, its price will have risen to 300 USD, and by the time it arrives in Antarctica, it will set you back 5,000 USD. ALE do try and help genuine expeditions by keeping costs as low as possible, but it is still a business, trying to make a profit in the most hostile environment on earth.

In 2010, with a year to go to the expedition, I started a new job in the operations department of Hereford. It was a two-year role, which also included a tour of duty in Afghanistan where I would help to run the operations room. Unfortunately for me, there was a clash of dates. My return date from Afghanistan was after the expedition was due to start, and my request for a change of dates was met with little sympathy from the SAS hierarchy. Your problem, you resolve it, was the message.

The only way of being able to make the dates for the expedition was to appeal to the good nature of the guy replacing me and see if he was prepared to come out to Afghanistan a month early, a big ask for a man with a family. Thankfully he agreed, and my fear of missing the expedition subsided.

Over the weeks and months that followed, Henry only managed to get us all together for a few training sessions where he would update us on developments. The teams were also confirmed, and I found out that I would be with Henry and he would be taking the Amundsen route. I began reading everything I could about Amundsen and Scott, and why one expedition failed and the other succeeded. It became almost like an obsession, and my home office became known as 'the Polar Office' as I began to gradually acquire books, maps and various pieces of equipment that I believed would be necessary for the journey.

Given everyone's busy professional lives, there was only enough time to practise everything we had learnt with a single ten-day training session in Norway.

Through various regimental connections, a Norwegian officer based in Norway, about two hours north of Oslo, agreed to host the entire team. It was an ideal opportunity to purchase some expedition equipment, including cross-country skis and Alfa expedition ski boots from Sportsnett, a specialist polar sports shop in Oslo. While the skis, bindings and poles had been pre-ordered, the boots hadn't, and I discovered to my horror that the shop didn't have any suitable boots in my size. It was a potentially disastrous development as we were heading straight out on expedition. But a crisis was averted when one of the sales assistants said that he had a pair of expedition boots in my size at home, which he was willing to sell me for £100. He rushed home, retrieved the boots and the deal was done. I felt that I had managed to get a bargain, although I was slightly concerned that they were not really what I was after. But there was nothing to be done about it and so no point in worrying. Sportsnett was an Aladdin's Cave of polar equipment. It somehow had a piece of equipment for every possible polar eventuality (apart from said boots) and, like children in a sweetshop, we found much of the kit irresistible, all spending far more than we intended.

Back at the base, we spent the first two days getting to grips with our equipment, such as testing our MSR stoves and ensuring we knew how to service them and conduct basic repairs, packing and repacking our pulks in a desperate bid to reduce weight, and attempting to learn the art of cross-country skiing. Henry was insistent that we established the correct skiing technique, which he said would be crucial if the expedition was to be successful. A failure to ski correctly, he said, would waste energy and could lead to damaged or broken skis and bindings, which could ultimately lead to the expedition being aborted.

Much to our surprise, Henry had previously explained that in terms of clothing, remarkably little was needed. The basic requirement was one set of thermal underwear, a couple of pairs of socks, a warm fleece for the evening, gloves and mittens, a thin windproof jacket for skiing and hauling, and a down jacket for when we had finished for the day.

The key pieces of equipment were goggles, skis and gloves, and we would require a spare pair of each, along with repair kits for our boots and bindings. We were also advised to buy the best sleeping bag we could afford. Each of us also had a closed-cell foam sleeping mat and some down booties to wear in the tent. The shared equipment included the tent (a four-man Hilleberg Keron GT four-season tunnel tent, the gold standard for polar expeditions), communications equipment, cooking fuel, and the crevasse rescue equipment, which consisted of ropes, ice screws and harnesses.

After two days of sorting the gear, the two teams drove north to begin training on Lake Femuden, where we would spend the next five days. The lake is the second-longest in Norway and freezes to a thickness of 30 feet in winter. It should have provided the perfect training platform, but a large dump of snow had fallen a few days before we arrived and it soon became clear that our progress would be severely limited.

Every team member had spent as much time as possible building up their stamina in preparation for the pulk-hauling by dragging Land Rover tyres around the English and Welsh countryside, but nothing had prepared us for dragging pulks through deep, soft snow. Within minutes I was out of breath, as we all were, and I felt my thigh muscles burn and cramp through a build-up of lactic acid. It was as if we were dragging overladen bath tubs through fields of sticky treacle.

Henry offered his usual words of encouragement and explained that it would be easier in Antarctica. But by the end of the first evening, we had progressed little more than a couple of miles. Over the course of the five days, the two teams hauled the pulks around 15 miles up to the top of the lake and then returned.

The effort required was intense and unrelenting, but all of us grew into the challenge and each day became slightly easier. The trip was also a perfect opportunity to get to know Henry and Lenny, the third man in our team, a bit better. Lenny had been in the SAS for several years. He was a former Para of Irish descent and was as tough as old boots. He had a great sense of humour and was fantastic for morale. The three of us got on perfectly well and I had the warm feeling that we would work really well together as a team when we eventually arrived in Antarctica.

On the last day, we came to a sort of group decision that we would put ourselves under a bit of physical pressure and ski and haul nine miles through the soft, fresh snow to our finish point.

It was a tough challenge, but we were all coping reasonably well for the first couple of hours, when it became clear that Lenny was struggling. It was almost as if his energy supply was failing and, step by step, he slipped further behind the rest of the team. As seasoned SAS men we were all aware that this can happen – even on a march. You go through an energy dip for an hour or so and then your body settles and you recover.

Henry was reluctant to slow the pace and it was Mark, the leader of the Scott team, who dropped back to help Lenny to try and establish what the problem was. Mark was also slightly miffed by Henry's attitude to his own teammate, and felt he should be the one dealing with Lenny.

Eventually Lenny's condition deteriorated to such an extent that we were forced to stop so that he could get some treatment.

'What's up, Lenny?' Henry enquired, his tone completely lacking in sympathy.

'Not sure, boss. Just struggling with the hauling. My energy levels feel very low. I've been finding the tab hard but manageable – like everyone else – until today.'

'Right, let's get him sorted. He needs fluids and some hot food then we'll make a call on what to do,' Henry advised.

Within an hour or so the colour had returned to Lenny's face, and he felt strong enough to continue, although he finished the day physically shattered. I wasn't particularly worried about him, I just assumed he was having a bad day, but I could see that Henry was concerned, although he kept his thoughts to himself.

The Norway expedition was a good shake-out for all of us. It helped with team bonding and it provided both teams with a taste of the sort of fastball the Antarctic might throw at us. There were lessons to be learnt and skills to be improved upon – exactly what we wanted. Lenny obviously realized there was a fitness issue, which he promised to resolve, but Henry also struggled with cold feet and I was aware that hauling the pulks was going to be a major challenge, more so than I had originally thought. The following day we flew back to the UK and settled back into our daily busy lives, rarely seeing each other and communicating only by email or phone.

Six months later, in June 2011, I boarded a plane to Afghanistan. It was approaching the height of summer and the base was like a small, dusty town, occupied by men and women in uniform and a few civilians. There was a boardwalk where you could buy pizza or Dunkin' Donuts, a KFC, several coffee houses, a Thai restaurant and a TGI Fridays, along with the various military canteens where troops from the NATO multinational force could eat. There were also various events and concerts to keep the troops entertained. In the winter it was often cold and wet, but by the summer the temperature could easily top 45°C and the whole base became a huge dustbowl. It was in operation 24/7, and the din of combat jets, attack helicopters and drones was ever present. But, despite being surrounded by some of the most powerful military hardware on the planet, it was sometimes easy to forget there was a war on, especially when you were stopped by the US Military Police for driving too fast.

Occasionally the Taliban would fire a few poorly aimed rockets, and on one occasion – as I walked back from the gym to the ops room – one landed just a few yards away. The rocket, which had been home-made or improvised, exploded with a huge crash and covered me in dust and gravel, but the shrapnel, which peppered some of the buildings inside the compound I was crossing, managed to miss me entirely. It was a pretty close call, and a reminder that in Afghanistan death was never far away. Other than that one incident, life inside the base was about as safe as you could be in Afghanistan.

Antarctica and the expedition soon slipped from my mind as I became embroiled in the running of operations. Days drifted by and weeks began to merge together. I was spending up to fourteen hours a day in the operations room, staring at computer screens and monitoring the activities of patrols around southern

Afghanistan. I would much rather have been out on the ground, but my job as an operations officer effectively meant I was desk-bound and I had to force myself to find time to hit the gym or put in several hours of tyre hauling, which put a lot of stress on the legs. My route took me around the base for two hours at a time, either in the morning or evening, when the temperature dropped from a high of 40°C to around 30°C. My tyre-dragging efforts attracted quite a lot of bemusement and wisecracks, especially from the US troops. 'Looks like a bit of a drag, buddy' was a favourite at the time, as well as: 'You must be really "tyred".'

When I eventually did arrive in the UK after the Afghan tour, I had just five days before I had to fly out to Chile – barely enough time to reacclimatize to the English weather, let alone the Antarctic. I was so busy that the fact that I was no longer in a war zone barely registered.

Time was so short at home that I insisted that Lucy flew with me and the other team members to Punta Arenas, in the southernmost tip of Chile, near Cape Horn.

Our base in Punta Arenas was the ALE warehouse, where a few frantic days were spent preparing the rations and packing the equipment, trying to whittle down the weight whilst at the same time ensuring that nothing urgent was left out. The white gas cooking fuel we had reserved would be collected in Antarctica and shared out between the teams, then packed into the pulks so that if the containers did rupture, our food would not be contaminated.

Every electrical item, such as satphones (satellite phones), GPS, cameras and our solar panels, which we would use to recharge batteries, was checked and rechecked. I forced myself to justify the need for every piece of equipment, and spent hours packing and repacking my pulk in a desperate and ultimately unsuccessful attempt to reduce the weight. Our rations were composed of dehydrated freeze-dried meals – along with energy drinks and grazing bags full of dried fruit, chocolate, nuts and seeds.

The one crucial piece of equipment we were told not to forget was a pee bottle. Wandering off into a whiteout for a midnight wee in temperatures of -30°C was not recommended.

While I was working on my kit, Lucy disappeared off to a local supermarket to buy cheese and salami for my grazing bag, stuff I could dip into between meal stops. We said our goodbyes at the end of the first week and I told her not to worry – my standard departure line whenever I left for some dangerous place around the world. I had said the same thing on numerous occasions. But leaving for Antarctica was different. I felt elated. There were no terrorists or warlords to fight, no one would be shooting at me; just 900 miles of emptiness to ski across.

An hour or so after Lucy departed, ALE informed us that our flight out to Antarctica was delayed due to bad weather. By the time we eventually left Chile, we were seven days behind schedule.

The team flew out on a giant chartered Ilyushin transport aircraft, a lumbering four-engine beast that easily swallowed all six of us, plus our equipment, and another forty or so passengers and cargo. There were a few other expeditioners on board, but most of our fellow travellers were scientists and ALE staff.

Four hours later, the Ilyushin touched down on the near-two-mile-long, diamond-hard, pristine blue ice runway at Union Glacier, known as UG in the polar world. I was almost overwhelmed by a sense of excitement and emotion. As I climbed down the aircraft stairs onto the ice below, feeling a little like Neil Armstrong stepping out of the *Eagle*, a member of the ALE team said, 'Welcome to Antarctica.' I reached out to shake her hand, slipped, and fell straight over onto my backside.

7

SOUTH

Adventure is just bad planning.

Roald amundsen

Antarctica was everything I had expected, and more. It was pure, raw, unspoilt, and looked like an Arabian desert where sand dunes had been replaced by endless snow and ice – a surface that can best be described as something like a vast lemon meringue pie. The sun shone as brightly as on an English summer's day, but a beast of a wind swept across the glacier and sent the temperature plunging to -25°C. The air was clean and devoid of any identifiable scent. All the other remote places on earth that I have visited as a soldier have a distinctive smell – the dustiness of Afghanistan, the sweaty humidity of Iraq, and the fetid, dank reek of decomposition in the jungle. But not Antarctica.

I knelt down and touched the ice, feeling an odd connection, almost as if this meeting was preordained. It is a strange experience to finally be confronted with something you have waited all your life to experience. I was happier and more excited by the prospect of a true adventure than I had ever been.

'Special, isn't it?' Henry said.

'Stunning,' I said. 'I expected it to be spectacular but this is breathtaking. You really do feel like you have arrived in another world.'

'We have,' added Lenny, smiling and walking away from the aircraft.

'Come on,' said Mark, 'Time for happy snaps later. We need to get our kit sorted and move across to the camp. There will be a briefing in an hour or so.'

ALE had created a small tented camp on the glacier and provided us with meals and hot drinks while we waited for a window of good weather that would finally allow us to fly to the start point. Both teams slept in their own tents at night but spent the day up at the main cookhouse tent relaxing, reading, playing a few board games, and eating as much fresh food as possible.

ALE was only allowed to establish a base at UG on the proviso that the area was not contaminated in any way, in line with all the environmental permits they held. This meant that no waste, including human, could be buried at the site. All waste water, including urine and faeces, was collected and contained in vast holding tanks, which were then flown out at the end of the season. When the site was closed down, there was absolutely no sign whatsoever that there had been human habitation there.

It was another five days before the team was given the green light. We boarded an aging DC3 Basler aircraft at 1300 hours on 2 November 2011 for a long, bumpy, seven-hour flight to the Bay of Whales. The crew had to stop halfway to refuel on the ice, where fuel depots had been established earlier in the season. But by 9 p.m. that evening we were at the start point, at the approximate location from where Amundsen and his team had originally established their camp Framheim before setting off a century earlier. Unlike Scott's hut, which had been built on solid ground, Amundsen's hut had been on the edge of a floating ice shelf that had long since carved off into the Southern Ocean. I felt, as we all did, that I was reaching back and touching history.

Under the gentlemen's agreement of the expedition, we had to wait until the other team was ready at their own start point, which was Scott's Hut, some 400 miles further along the edge of the Ross Ice Shelf. Incredibly, the hut is still in place today, almost untouched from the moment Scott and his party left for their tragic journey to the South Pole. Boxes of rations, cans of unopened food, and even reindeer-skin sleeping bags were still hanging on the same washing lines inside the hut.

'Hi Henry, we've arrived safely,' Mark said, calling Henry on the satphone. 'Had a quick look around Scott's hut. Incredible. It was as if they were there yesterday. The air is so dry here that everything has been preserved. Anyway. We are ready. Do you want to do the countdown?'

'Roger, will do,' added Henry. 'Glad you arrived safely. Wish you the best of luck and see you in a couple of months.'

After a pause for five seconds or so, Henry then said, 'On your marks. Go!'

And with that we were off. After all the months of planning and training and the anxiety over whether the expedition would go ahead, we were heading south on the adventure of a lifetime. I could not have felt any happier if I had won the lottery.

But the sheer joy of knowing that I was following in the footsteps of Amundsen was remarkably short-lived. Almost from the very first few hours of the expedition, Lenny began to fall behind; he was clearly struggling to keep up with the pace set by Henry. Immediately, I began to think of the last day in Norway, ten months ago, when Lenny had problems on the frozen lake. Surely, I thought to myself, this can't be mental. Lenny was a fit guy, a sergeant in the SAS. He used to be a boxer and I knew that he had been working hard on his fitness, but something was clearly amiss. I dropped backed and skied alongside him. He was sweating heavily and his head had dropped. His skin was waxy-white and he was almost gasping for air. I had seen the look before both on SAS selection and during my Royal Marines training. It is the look of someone who is just about to drop. What left me so confused was the fact that we hadn't been on the move for very long, maybe only a couple of hours, and we

Above: Out in the desert preparing for operations. Sleeping under the vehicles for shade during the heat of the day and only moving at night.

Below: Counter-narcotics operations, grenade in hand, ready to destroy an opium press.

Top: Henry and I negotiating massive crevassing on the Axel Heiberg Glacier, just as Amundsen had done a hundred years before us.

Left: Henry alongside Amundsen's Cairn, which has stood undisturbed for a century.

Above: Henry looking very happy on Christmas Day with the presents I surprised him with.

Left: The temperature dropped below - 30°C on the polar plateau;
I got frost nip in the tips of my fingers later that day.

Right: Holding the regimental flag signed by HRH Prince William
after we reached the South Pole.

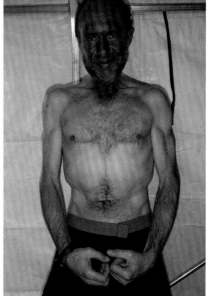

Above: 'I love it when a plan comes together':
I'm not a smoker but it seemed a fitting way to
celebrate the end of a successful expedition.

Right: Dramatic weight loss on reaching the
South Pole. We learnt a huge amount about
nutrition from that first trip.

Left: Alex practising crevasse rescue drills in Iceland before SPEAR17.

Below: The SPEAR17 team at Hercules Inlet on the coast of Antarctica, about to embark on our 1,100-mile crossing. *From left:* me, Alun, Ollie, Chris, Alex and James.

Above: Leading the team through some light sastrugi on our way towards the polar plateau.

Below: Alex and James taking a snow bath in -20°C. They only did this once!

Above: Re-enacting a Shackleton tradition, the Party in the Ritz – a mini-celebration every time we crossed a degree of latitude.

Left: Despite being cautious and covering up, Ollie's cold injuries on his lips caused him a lot of grief.

Right: The Swithinbank moraine field in the Shackleton Glacier. Huge boulders are transported along with the flow of ice.

Above: A parhelion, or 'sun dog', caused by refracted light from ice crystals in the atmosphere.

Below: Henry's memorial cairn at the top of the Shackleton Glacier, with stunning views back over the polar plateau.

Above: The team after we arrived at the South Pole on Christmas Day 2016.

Below: A jumble of disturbed ice and crevassing in the Shackleton Glacier hampered progress.

had made sure that we had all eaten and were properly hydrated before we departed.

'How you doing, mate?' I asked.

Lenny lifted his head: 'I'm OK. Just finding it a bit tougher than I thought.'

I stayed with Lenny for a while, trying to chat, but he was uncommunicative and we were slipping further behind with every step.

Eventually I quickened my pace and skied up to Henry.

'Lenny's struggling. He can't keep up with the pace and I'm worried about him. I don't think it is just his fitness. We need to get him in the tent. We're going to have to call it a day.'

Henry stopped skiing, said OK, and began unpacking his pulk. His silence spoke volumes and I could tell that he was frustrated.

'Sorry, fellas,' Lenny said as he skied towards where Henry and I had begun erecting the tent. Lenny looked exhausted and was really suffering. 'I don't know what's wrong with me. I was working my arse off but I just couldn't keep up with the pace you two were going. Maybe I've picked up a bug or something.'

As Lenny apologized profusely, I could tell he was embarrassed and felt as though he was letting us down. For any member of the SAS to struggle with a physical task is nothing short of a humiliation.

'Forget it,' Henry said. 'We're all going to have bad days – yours just happened to be today. You'll probably be fine tomorrow. Just rest tonight and get a good night's sleep.'

The atmosphere inside the tent that first night can best be described as solemn. The initial jubilation we all felt at the start point had disappeared as if it had never existed. Henry kept himself busy to avoid conversation, while I began the laborious task of melting snow. I filled a five-litre kettle with snow and boiled it down until it all had melted, leaving just an inch of water. This was because the ice and snow in Antarctica contains very little moisture. Once the snow had melted, I began the process again. It took a good two hours to boil enough water to make our energy drinks and rehydrate our meals.

Meanwhile, Henry wrote up his diary and prepared for the next day's events.

My diary entry for the first day read:

2 November. Weather conditions very good. About -15°C. No wind, sun shining. Only managed 1.96 nm [nautical miles] before Lenny began to struggle. Had to stop and put up tent. Very frustrated. Going was good, surface was very firm. Now switched to New Zealand time.

The weather was kind to us as we awoke to a perfectly blue, cloudless sky. It was about -5°C inside. Despite the cold, Henry was up seconds after his alarm clock rang into life.

'Let's get going, chaps. We've a lot to do today,' Henry said, urging us out of our sleeping bags.

I glanced across at Lenny, whose movement was slow and laboured. He already looked exhausted, but I also knew that the nature of his character meant that he would be trying to make amends for his poor start to the expedition.

Breakfast was porridge and a hot drink, and then we were out, taking down the tent and packing our pulks. By the time we were ready to start moving, we were all cold. Our down-filled duvet jackets had been ditched for a single-layer, thin windproof jacket, so that we wouldn't get too hot, but for the next thirty minutes or so we would freeze until our bodies warmed with the effort of hauling.

Almost immediately I noticed Lenny slipping behind. Henry noticed too but refused to slow his pace, and pressed on with me following close behind. Eventually I told Henry that we had to stop.

'We have a problem with Lenny,' I said, removing my goggles and staring back into the distance.

'I know. Any clues as to what is going on with him? Do you think his heart is in this?' Henry said, gesticulating towards Lenny who was a dot in the distance about half a mile away.

'Yes, it is. He wouldn't have gone through this whole process if he wasn't up for it.'

Thirty minutes later Lenny arrived. He was clearly in discomfort. Physically he was worse than the previous day but, apart from struggling with pulling the pulk, there were no symptoms of any other illness.

Henry agreed to rest for around ten minutes, but we soon began to feel the effects of the cold and we set off again at a pace Henry and I thought Lenny would be able to cope with. Once again, though, he quickly fell behind. Again, I called time and told Henry we needed to stop – a fact to which he was already resigned. When Lenny caught up with us it was clear that he couldn't continue – he had effectively become a casualty. By the time he reached us, his face was a deathly white and he was dry retching.

My diary entry for the second day sums up the situation.

3 November. Weather very bright. Set off and quickly became evident that Lenny was struggling. He dropped back after twenty minutes. Tried to encourage him but was exhausted. Henry took the decision for me and him to take 20 kg each off Lenny. Did this but found extra weight really difficult. Didn't make much difference to Lenny. When we stopped again he looked pretty white and started vomiting. I told Henry we need to get a tent up ASAP. Got Lenny into dry clothes and sleeping bag. Really bored for rest of day. 1.5 nm covered. Wonder how Scott team are doing?

It was the worst of all possible starts. We had covered just over three miles and were now facing serious problems. Henry told me privately that there was no way Lenny could continue, but he wanted Lenny to reach that decision himself. As we melted snow for some warm drinks, Henry broached the subject with Lenny.

'Do you think you can carry on?' Henry asked. I thought his approach lacked sympathy but that was Henry's way. He was a kind, gentle and thoughtful man who showed great empathy for friends and colleagues. But on expedition he was far more driven.

'I don't think I can, boss,' Lenny answered. 'I feel terrible, and I can't see anything which is going to change that. I'm really struggling to walk, let alone pull all that weight. I think I'm going to have to pull out.'

It was not the answer that Henry wanted to hear, but two minutes later he was on the satphone calling the ALE operations room requesting an emergency extraction for Lenny.

Henry returned to the tent and looked at me and said: 'No move tomorrow, the Basler will be another forty-eight hours. We will stay put tomorrow and push on once Lenny has been extracted.'

Henry spoke as if Lenny wasn't in the tent. As far as he was concerned, Lenny was a problem that needed to be resolved – nothing more.

'Tough luck, mate,' I said to Lenny, who nodded and tried to smile.

'I just can't apologize enough,' Lenny added. 'I'm so sorry. I feel like I've let everyone down.'

'Don't worry about it, mate. Not your fault. Just get yourself properly checked out when you get home. Maybe you do have a virus,' I added, but realized my attempt at sympathy was lost on Lenny.

There was little chat that evening. Everyone felt deflated, no one more so than Lenny. The following day was mostly spent in the tent listening to music and playing cards. Lenny's condition appeared to stabilize, but he was still ill and it was the correct call to get him extracted.

Henry and I busied ourselves by repacking the shared equipment – fuel, spares, medical pack and the crevasse rescue equipment – between the two of us. Our pulks must have tipped the scales at around 130 kilograms, around twice our own body weight. The extra equipment increased the weight of our pulks by 10 kilograms. It was unwanted burden, but there was little we could do about it.

By the time Lenny was eventually extracted, it was the early afternoon of 6 November; we had jumped ahead a day because we had transferred to New Zealand time. Of all the things we had expected to happen, an early departure from the team was the last of them. Henry and I wished Lenny the best of luck and then got on our way, knowing that we had to make up a lot of lost time.

We both watched as the aircraft safely took off and headed for Union Glacier. To be honest, I was relieved that Lenny had gone, because his poor health had just become a burden. If we had been on an SAS operation, somewhere in Iraq or Afghanistan, Lenny would have been casevaced without a moment's hesitation, and we had to adopt the same, hard-nosed mentality. It wasn't personal, and I would have been treated in the same way if it had been me who had gone down rather than Lenny.

As soon as the Basler BT-67 was airborne, we set off in earnest, Henry leading for the first hour then me up ahead. It felt good to get some miles under our skis and I felt my spirits lifting almost immediately. We covered just another four miles that afternoon, and in the evening we wrote up our diaries and phoned through our audio message informing our expedition media manager that Lenny had left the expedition.

My diary entry read:

6 November. Chilled in the morning while waiting for the Basler for Lenny. Was two hours late. Came in at 1330 hours. Spent ages circling overhead to find a suitable landing site as we were in a crevasse field. Eventually landed. Looked dodgy. Loaded Lenny in. Five minutes and off they went. Just me and Henry now.

The following morning we were up early and desperate to get on the move and get back on schedule, almost racing each other in a bid to make up the lost time. In front of us lay 450 miles of the Ross Ice Shelf and at least six weeks of physically hard and mentally challenging hauling until we reached the Transantarctic Mountains and the Axel Heiberg Glacier. The day's skiing finished around 7 p.m. It felt good to have completed a day without all the stopping and starting that had caused so much frustration for the first two days of the expedition.

Routine is key to Antarctic travel. By the end of the evening you are often too tired to think, so being able to operate on autopilot – which allows for unspoken teamwork – is crucial. The first and most important job was to pitch the tent as quickly as possible. Pulling a sledge across ice is hard work, and the body generates enough heat to keep you warm, even in the summer when the temperature can reach -40°C. But as soon as you stop, the body temperature drops like a stone.

Our evening routine was amended slightly now that Lenny had gone. Henry was responsible for pitching the tent whilst my job was to prepare food and drinks.

With the tent up, getting the cooker on was always the main priority. Henry would secure the pulks, then would begin digging a footwell in the vestibule to make it easy to remove our boots, while I set about melting snow. My meal routine was to make hot energy drinks first of all, then boil several more litres to rehydrate our meals and hot puddings and to provide us with some water for the morning and the next day. It was a laborious process that took several hours.

As well as the water preparation, there was a seemingly endless list of jobs that needed attention, including equipment repairs, sewing up tears in trousers and jackets, regluing soles onto boots, writing our diaries, delivering the audio blog and making satphone calls. It was often midnight before we were ready for bed.

The first night after Lenny left us, Henry switched on the satphone and began phoning in the audio message for the expedition blog. Back in the UK we had an expedition media manager and we would leave an audio message on his voicemail, which he would then transcribe and post on the expedition website, along with our daily stats of mileage covered, weather conditions, etc. Henry always finished off his blogs with the salutation, 'Onwards', a tradition I've carried on to this day.

He also received a message that the Army were concerned about the fact that there were now only two of us, which impacted the risk assessment.

'They suggest we *abort*,' Henry told me.

I was stunned into silence. This was a big deal, and as expedition leader it was ultimately Henry's call. The problem for us was that in Henry's risk assessment, he had stated that the expedition could not continue if a team was reduced to two men because of the problem of crevasses. It could be extremely difficult for Henry to rescue me on his own if I fell into a crevasse with my pulk, or vice versa.

'Well, I'm not open to their suggestion,' Henry continued. 'Are you happy to continue, Lou? I know it's a big call, but if it all goes wrong this is on me, I'll take the rap. Anyway, what are they going to do? Come and get us?'

'Yep, let's continue,' I said. 'I don't want to give up now.'

Henry and I both came to the conclusion that if the expedition was successful no one would complain and if it turned into a disaster the fact that we broke the rules would be the least of our problems. But privately I wondered whether the expedition was cursed. Firstly, the weather delays, then Lenny, and now this.

As the days passed by, we got used to the vagaries of the Antarctic weather, which could change almost hour by hour. There were days when the weather closed in and the visibility decreased so much that I felt as if we were making little progress, but our GPS device said otherwise. Skiing when the sun shone brightly and the weather was clear – even if it was into a strong and tiring headwind – was always more pleasant than on a dull, sunless day.

After around three weeks, when we had managed to cross around half the Ross Ice Shelf, and just when I thought I was getting to grips with the weather, the ice and the cold, I began to develop serious blisters. I was beginning to pay the price of buying the wrong boots in Norway at a knockdown price of £100. What I thought was such a great bargain was now proving very costly.

The issue was caused by deep, soft snow, which meant that to get the pulk moving required much more effort. I was forced to pull with my body at an angle of almost 45 degrees and there was a huge amount of pressure on the backs of my heels as they were jammed into the back of the boots. In really soft snow, it was almost impossible to gain any forward momentum, and my progress was reduced to single, exhausting steps. It was insanely hard work and the lactic acid build-up in my legs meant that I had to stop and rest every minute or so. As I drove forward, the rubbing in the back of the boots began to strip the skin from my heels. With every step the rubbing intensified, and I gradually fell behind Henry, unable to keep up or take my turn in breaking trail.

On one day I watched as Henry seemed to glide effortlessly into the distance, becoming nothing more than a tiny speck on the horizon. Henry refused to slow down, and rightly so. He had established a mental daily schedule of the distance we needed to cover if we weren't going to run out of food. Even though we had both been skiing for up to ten hours a day, sometimes more, we still hadn't made up the time lost at the beginning of the expedition.

The issue with the boots and my lacerated heels was one of the hardest parts of the expedition for me. It was constant and unrelenting pain for hour upon hour, day after day, week after week. The blisters had a major impact on my morale because I couldn't haul or ski without being in some kind of pain.

I came to the end of my tether one evening when I arrived at our camping spot. Henry had already erected the tent by the time I had arrived and was in the process of melting snow in the kettle. I was slightly cross that he hadn't waited for me to catch up, but I was more annoyed with myself for not being able to keep up with the pace. Henry handed me a hot drink when I entered the tent but that didn't improve my mood.

'I can't believe you've come back here for a second time,' I said to Henry gloomily. 'This is so horrendous – I'm never coming back here. It's just too much. Every part of my body aches, I'm holding you back and my feet are falling apart.'

The joy I had felt when I first arrived had become a faded dream, and I seriously questioned for the first time what I was doing in the Antarctic. Henry's words two years earlier when he told of the difficulties of daily life on the continent were ringing in my ears.

Henry chuckled: 'You have to learn to work with the environment, not fight against it. Accept that we are incredibly lucky and privileged to be here. This is an amazing place. What we are doing is amazing. Sure, your feet are giving you grief, but you'll get over it and it will seem like a minor inconvenience.'

But I just couldn't see it. I winced in pain as I removed my boots and, as I peeled off my liner socks, huge flaps of skin peeled off with them, exposing raw flesh below.

Henry glanced at my feet. 'That looks grim,' he said, adding more snow into the kettle. By now I knew that for some reason Henry lacked any real sympathy on expedition – it was all part of his single-minded determination to achieve the goal.

I let my feet air for as long as possible that night, and seriously wondered how much further I could ski. If my blisters became infected, my expedition was over. I taped my feet as heavily as I could the following morning but, within an hour of starting, I could barely walk. Henry was already way ahead. I had to take some very drastic action. I stopped, opened up my pulk, retrieved my Leatherman knife and began sawing and cutting away at the back of my boots, opening up a square hole so that the pressure on my heels was relieved. It was the last roll of the dice as far as I was concerned.

I set off again and felt some of the pressure ease. The relief was instantaneous, and I was soon skiing at my normal pace. I was nervous about the open backs, the potential impact on the structure of the boot, as well as the cold getting in, but I had a plan to resolve that issue. By lunchtime I had caught up with Henry and showed him my handiwork.

'Ingenious,' he said, smiling. 'You know Ranulph Fiennes had the same problem and cut holes in his boots as well. You're in good company.'

'So relieved,' I responded. 'This is the first time in about ten days that I haven't felt in serious pain.'

That evening I glued some material from the gaiter to the inside of the boot, just to keep the snow and spindrift out and to prevent my feet from becoming too cold. Remarkably, it worked. Then, slowly but surely, with the pressure on my heels eased, the large open areas of raw flesh began to heal and new, harder skin began to form.

I wasn't the only one to be a victim of an Antarctic mishap. About midway to the Pole, while crossing the Ross Ice Shelf, Henry delved into his grazing bag, which for both of us consisted of a few nuts, chocolate, cheese, salami and boiled sweets. Chocolate would often freeze to a hardness resembling something like concrete, so the rule was to allow the warmth of your mouth to soften it before biting into it. Unfortunately, Henry forgot to do this during one of our five-minute breaks, and the rock-hard chocolate snapped a crown, exposing the nerve endings of one of his front teeth.

The -30°C supercooled air that whistled into Henry's mouth whenever he breathed in sent the exposed nerves into a violent spasm, but there was little we could do until that evening, so he just had to try and cope the best he could. As we skied to our camp that night, Henry would occasionally stop, clearly in pain and holding his jaw. But he never once complained. His stoicism was truly impressive. Henry was famously short on sympathy when he was on expedition, but he never asked for any either when he was suffering.

Fortunately, part of the expedition training included a few basic lessons in field dentistry, and I was able to relieve the pain by packing some special dental cement into the exposed cavity.

Despite the injuries, the close confines of sharing a tent and the total absence of all privacy, there was only one moment of tension during the entire expedition. The nearest thing we had to a stand-up row came at the end of one particularly gruelling day on the Ross Ice Shelf.

Mornings were always the worst part of any day. Neither Henry nor I ever felt we had slept for long enough. Both of us always woke feeling exhausted and unwilling to confront the misery of leaving a warm sleeping bag for the cold reality of an Antarctic morning, where even on a good day the temperature inside the tent might be as low as -5°C. We chatted little as we went through the process of boiling water, making breakfast and packing up our equipment. The tent was the last item to be dismantled, and then came the worst moment of all, when we would take off and pack away our warm down jackets prior to setting off. Both of us skied in a thin, windproof jacket, a long-sleeved thermal vest and trousers. This was deliberate, so that we did not get overheated and sweat. In subzero temperatures, sweat freezes on the inside of clothing, causing layers to lose their thermal efficiency and resulting in the rapid cooling of a body's core temperature. This could and often did lead to hypothermia. I had seen this happen on numerous occasions during my time in the SAS, most notably during a six-week exercise in Norway, when one young and relatively inexperienced member of my troop had to be evacuated to hospital after developing hypothermia.

Wearing minimal garments for skiing, though, meant the period between stripping off our cosy jackets and getting warm – which on a very cold day could take at least an hour – was miserable. This was why we both dreaded mornings.

The day in question had been one of those when nothing seems to go right, when from the very first moment you feel as though you should have stayed in bed. I was struggling hauling my pulk – the snow and ice seemed to be particularly soft – and, try as I might, I couldn't keep pace with Henry. No words were exchanged during the breaks, I was just too tired to talk, and by the end of the day I was shattered. I caught up with Henry and he gave me one of those 'what took you so long?' looks.

'I'm a bit disappointed with what we've done so far today. We've only managed fourteen nautical miles. Do you reckon we can push a bit harder on this final leg while you're up front? If not, we might have to ski for a bit longer this evening.'

I couldn't believe what Henry had just said. I was completely exhausted, and he seemed to be implying that I wasn't trying hard enough. The sudden surge of anger reinvigorated me.

'OK fine. Let's go.'

I set off like a man possessed, skiing faster than I had ever done before and determined not to look behind me for the whole seventy-five minutes.

'I'll show him,' I said to myself. 'He'll regret this. He's going to be hanging out.'

I pushed harder and harder, breaking all the rules, and sweating profusely. At times I was almost running across the snow. I was absolutely determined to make Henry feel as bad as I had done so many times when I was struggling with my blisters and he had left me trailing in his wake.

I checked my watch – half an hour had passed but the anger still raged inside me. I was determined to stay with the pace, no matter what the cost. At the end of this leg I wanted to see Henry struggling, nothing but a distant, floundering dot on the horizon.

With the seventy-five minutes up, I stopped. I was exhausted and could barely stand. The frost and frozen moisture was hanging off my face, my clothes were soaked with sweat but it had been worth it – until I turned around. Henry was right behind me! But he was completely hanging out, bent over double on his ski poles, gasping for breath, with an expression of abject pain on his face. He looked much older than his fifty-one years. I nearly fell over in shock and my anger disappeared. I was seriously impressed. I had tried my best to blast him away, but he had stayed with me and without a word of complaint.

I took off my glove and showed Henry my hand: 'Respect,' I said.

'What?' he responded, looking confused. 'What are you talking about? What's going on? I don't know why, but I really struggled on that last leg.'

'I had a bit of a sad-on when you said that I needed to push harder, so I decided I'd try and drop you way behind. That's why I was pushing so hard. I can't believe that you remained right on my shoulder.'

Henry leaned forward on his poles and began laughing. 'Thank God for that. I thought I was having a really bad moment.

I couldn't work out why I was struggling so badly. I didn't realize I was getting a beasting!'

From that point on I had a whole new level of respect and admiration for Henry and the deep reserves of physical and mental endurance he possessed. I remember thinking, 'This man is carved from granite.'

By the end of Day 37 we had reached the foot of the Transantarctic Mountains and I was exhausted. Each of the 12 nautical miles we skied that day had been a real struggle, almost as if some unknown force had been acting against us. The plan was to finish at 7 p.m., and for the last two hours I had been desperately clock-watching, urging on every minute.

With every step, my pulk seemed to become heavier, my steps more laboured. I sucked the cold air into my lungs and was barely able to lift my head. When Henry stopped and raised a pole to signify we had reached our overnight camp spot, I almost collapsed to my knees in relief.

Since leaving the Bay of Whales on 11 November, we had skied more than 450 miles, across the undulating crystalline Ross Ice Shelf to the foot of the majestic, snow-covered Transantarctic Mountains, a great natural barrier that spans the frozen continent.

'That's the Axel Heiberg glacier,' Henry said gleefully, pointing at a mass of diamond-white ice rising high into the mountains. 'Roald Amundsen was the first to ascend it one hundred years ago, and tomorrow we'll attempt the same.'

The glacier was our frozen gateway up onto the polar plateau, but its beguiling beauty hid many dangers. Dozens of concealed snow-covered crevasses ran up through the range, making the ascent potentially hazardous, especially with an 80-kilogram pulk in tow.

We had arrived in one of the remotest places on earth. Two inconspicuous specks of black on an otherwise snowy carpet of perfect white. It should have been a special moment, one where you were left awestruck by the minimalist beauty. But, to be honest, it had been a long, tiring day, and all I could think of was getting some hot food inside my belly and a good night's sleep.

As I refilled the kettle with snow, Henry came into the tent. 'Wind's picked up,' he said, flicking flakes of powdery snow from his down jacket.

8

THE CAIRN

*For scientific discovery give me Scott; for speed and
efficiency of travel give me Amundsen; but when
disaster strikes and all hope is gone, get down on
your knees and pray for Shackleton.*

RAYMOND PRIESTLEY

Henry's evening routine inside the tent always began with him reading from a copy of Amundsen's diary of his race to the Pole. In fact, both Henry and I had read dozens of books about the polar explorers and tried to learn from their successes and failures – our obsession was part of why we'd gelled so easily, despite our differences in rank and upbringing.

Unlike Scott, who planned to use motor sledges, dogs and ponies during his journey south, Amundsen trusted only in the use of dogs. On 19 October 1911, the Norwegian explorer departed his base camp with fifty-two dogs pulling four sledges, the remainder of his party skiing alongside. Along the route, Amundsen's team would slaughter the weaker dogs and feed them to the remainder of the team. It was a ruthless but efficient strategy that, quite simply, worked.

The dogs could only work for five to six hours a day and so Amundsen's team were still strong and well rested by the time they reached the Pole on 14 December, a month before Scott's five-man party arrived. Amundsen eventually returned to his base on 25 January 1912, with eleven surviving dogs. Just six days afterwards, Scott arrived at the Pole.

It has been widely reported that Amundsen succeeded where Scott had failed because he was better prepared, better equipped, and excelled in the use of dogs in polar conditions. Amundsen's sole aim was to be the first to get to the South Pole – it was the single focus of the entire expedition. By comparison with Scott, Amundsen's journey south was almost a breeze. His team started every day well rested and well fed, and he arrived at the Pole more or less on schedule.

Henry marvelled at Amundsen's efficiency, ruthlessness and single-minded approach to achieving the mission.

'It says here,' he said, pointing at a section of the diary, 'that Amundsen decided that he was going to lay down a half-ton depot of food, dog food and fuel for the return journey. We are at that position now.'

I listened with interest as I packed more snow into the kettle.

'I've researched this, and on the return journey he sent two of his team out to a nearby rocky outcrop, which Amundsen named Mount Betty after his childhood nurse back in Norway, to build a stone cairn.' Henry looked up at me to gauge my reaction. There was a mischievous, almost worrying, glint in his eye. I nodded and smiled and wondered why Henry was so excited.

He continued reading: 'The diary says that the main purpose of the cairn was to act as a navigational aid or marker and to leave a record of their expedition, should they not make it back. Amundsen told the pair to place a can of cooking fuel and a tin of matches inside the cairn to act as an emergency reserve. I reckon there's a chance it could still be there.'

Then the penny dropped, and I knew what was coming.

'Look, I really want to go out, see if I can locate Mount Betty and find this cairn. It's a fantastic opportunity.'

Henry's piercing blue eyes shone brightly out of his craggy, weather-beaten face, air-dried by the Antarctic wind and scarred by frostbite. He was a polar romantic. If he could have lived his life again, it would have been in the first decade of the twentieth century, as a member of a Scott or Shackleton expedition like his distant ancestor Frank Worsley. His enthusiasm was palpable, but there were real dangers. The cairn's exact location was vague, the weather was deteriorating, and we had just finished a long, hard day on the ice. I was exhausted, possibly dangerously so. The last thing I wanted to do was trek into the unknown for another four or five more hours. I also had an important equipment repair to sort out.

I felt terribly conflicted. I could see Henry wanted me to come along, and I wanted to support him. I completely understood why this was so important to him, but I also knew my own body and my limitations.

But he was undeterred: 'Lou, this is a once-in-a-lifetime opportunity, and I really want to do it. You stay here. I know it's not the best idea to go out on my own, but I reckon I've got a really good chance of finding the cairn.'

Henry was the boss. He was a lieutenant colonel in the SAS and the expedition leader. I was a warrant officer and in no position to argue.

'The last thing your wife said to me was not to allow you to go off and do anything stupid.' I reminded him of Joanna's words.

Henry laughed. 'It will be fine. If I'm not back in three hours, it probably means I've fallen down a crevasse. You come out after me. Bring the crevasse rescue kit and follow my ski tracks and you'll eventually find me. But don't worry, everything will be fine.'

A few minutes later, Henry was on his skis and had disappeared into the swirling, snow-filled wind. He had a flask with a hot drink, a GPS and his camera. At that very moment, I was reminded of the tragic scene immortalized in the 1948 film *Scott of the Antarctic*, when Captain Oates had headed out on his final, lonely journey. I also wondered if that was the last I would see of Henry, and whether allowing him to venture out on his own would prove to be a fatal miscalculation.

But he had gone and, rather than worry about it, I tried to distract myself by tackling the long list of jobs I needed to complete that evening. I made myself spaghetti bolognese and chocolate pudding, all the time trying to push away images of Henry slipping off the edge of an ice-covered cliff or plunging into a crevasse. Antarctic glaciers tend to be riddled with crevasses and, even though Henry was following a route up the Axel Heiberg's edge, he was still moving through a high-risk area.

The wind was picking up again and began to pound the tent with powerful gusts. The knot in my stomach tightened as I started on the repair to my harness. Three hours had passed since Henry had left the tent. I began to think through the steps I'd need to take if he didn't arrive soon.

About ten minutes later, I put my boots on and unzipped the tent, but could see nothing. The wind had whipped up the snow into a swirling mass, creating a whiteout. Visibility was intermittent and probably less than 20 yards. The agreed plan was that I should now implement the rescue procedure, which would normally require an urgent emergency phone call to ALE, our support providers. But what was I going to say? 'Hi, it's Lou. Bit of a problem my end. Henry's wandered off on his own and I haven't seen him for three hours. Oh, and the weather is deteriorating, rapidly.'

It didn't sound too clever, and a message like that could have serious repercussions for the expedition. I decided to give him a little longer, trying to convince myself it was the best course of action, but not before I had peered out of the tent again, hoping that I would see Henry emerging from the whiteout with that big friendly smile of his. But there was nothing.

I withdrew inside the tent, cocooned from the elements, and watched the minutes tick by. The wind now sounded like someone banging on the side of the tent with a large piece of wood, and it was getting louder and stronger. The situation didn't look good. Every five minutes or so I would unzip the main door and pop my head out, convinced that Henry would emerge, snow-covered and safe, but there was no sign of him.

At midnight, four hours after Henry had departed, the time for waiting was over. He had probably slipped into a crevasse and I prayed that he was still alive. The question was: should I alert ALE? Not yet, I decided. There was no need to panic. All my SAS training and experience dictated that what was needed was a calm head and a logical plan. I needed to establish some ground truth. There was simply no point in phoning ALE and saying Henry was missing and I don't where he is or what has happened to him. I needed at the very least to establish whether he was alive, and to get a fix on his approximate location.

I grabbed the satphone and first-aid kit. I packed the emergency rescue kit – 150 feet of rope, climbing harness, pulleys and ice screws – into the pulk, along with some warm kit, food and water. I was just fitting my skis when, through the swirling snow, I saw a figure emerging – it could only be Henry. My heart rate slowed. As Henry skied towards me, he began waving his arms frantically. I could see that he was shouting, but the buffeting wind was drowning out all sound.

Ice crystals dangled from around the mouthpiece of his black balaclava where the moisture in his breath had frozen. His skiing was slow and laboured and he looked exhausted. Eventually he came within shouting distance.

'I found it, I found it. I've only found the bloody cairn.'

Henry was ecstatic. For someone so obsessed with polar history, he couldn't have asked for more. I helped remove his skis and we quickly dived back inside the tent to warm him up.

I fired up the cooker and Henry pulled out his camera and showed me a video and stills of the cairn.

'It took me ages to find it,' he said, removing his gloves and jacket. 'I climbed up the edge of the Axel Heiberg through the whiteout. Once I got to a certain altitude, everything was clear and I had a fantastic view.' He was speaking so quickly he barely paused for breath. 'I could see the obvious rocky outcrop, fitting the description of Mount Betty, in exactly the place where Amundsen described it in his diary.'

I handed Henry a mug of over-sweetened hot chocolate and poured the rest of the hot water into his sachet of dehydrated spaghetti bolognese.

Henry nodded his thanks as he continued describing his adventure: 'It was a couple of miles away so I skied off towards it. Removed my skis, scrambled up the rock, and there on the top was this blatantly obvious manmade cairn. I couldn't believe that I'd found it. I carefully lifted a stone and peered inside and I could see this rusty can of cooking fuel with an equally rusty tin of matches on top in the same position as they were a hundred years ago. Imagine that – not moved an inch, despite everything the Antarctic has thrown at them. Unbelievable. I placed everything back so it looks completely undisturbed, maybe to rest untouched for another hundred years.'

The relief I felt knowing that Henry was safe was almost uncontainable, and I was incredibly excited for him – it was like watching someone finally feel as though they were complete.

'Come on, Lou, get dressed,' Henry said. 'Get your kit on and I'll take you. We can be there and back in three or four hours.'

I couldn't help but smile at Henry. This had been his moment, a defining point in his life. It was as though he had reached back in time and touched history, and I couldn't have been more pleased for him. But I felt the need to take a step back and look at the bigger picture. The challenge was to get to the Pole safe and sound, and in the end my common sense prevailed. We both decided to get a good night's sleep and stick to the plan.

The following morning we began the ascent of the Axel Heiberg, which ran up through the mountains like an ice super-highway, surrounded on all sides by dramatic and towering snow-capped peaks. It was like some alternative world, where Henry and I were the only two people on the planet. Apart from Amundsen's cairn, there was absolutely no sign that anyone else had ever ventured to this empty wilderness at the bottom of the world.

Henry set off trying to ski up the glacier – while dragging the pulk – through soft, deep snow, and I followed in his tracks. Skiing uphill is not easy at the best of times, but it verges on impossible when dragging an 80-kilogram dead weight.

After two hours of hard work with very little gain, Henry then tried to traverse up the slope, but again progress was too slow. No one had climbed the Axel Heiberg unsupported since Amundsen and his team one hundred years earlier, but he had had around ten dogs pulling each of his four sledges. By comparison his challenge was a breeze. The only option left was to double-haul each load.

I attached my harness to Henry's pulk, and we dragged and skied up the glacier for 300 yards or so, then stopped and skied back down to mine and repeated the process. To go forward one mile we were effectively skiing three miles, which turned the 30-mile trek into a demoralizing 90-mile undertaking.

Although hard work, it was the first time during the expedition that Henry and I could talk to each other while we were on the move. Our usual mode of travelling meant that we were always in single file, so the double-hauling meant that we could actually chat properly, as well as take in the extraordinary landscape. We stopped and took a few pictures of huge crevasses or strange ice formations, and beautiful unspoilt and largely unexplored mountain ranges. We cracked jokes and talked about the world and all its problems, what we planned to do after leaving the Army and what aspirations we had for our children. It was probably during those six days of skiing alongside each other, and the shared adversity, that I really got to know Henry and what made him tick.

As with most days, one of the hardest parts was at the end, when you were close to physical exhaustion. The slope of the glacier meant that on a couple of occasions we had to dig a flat shelf so that we had a base on which to pitch the tent. It would take up to ninety minutes of hard digging into the ice to create a stable platform. It was absolutely the last thing you wanted to do after climbing the glacier all day.

As we ascended the glacier, our progress was also slowed by the endless number of crevasses. Some we could simply step across, but others would force us into a detour of hundreds of yards before we could find a safe crossing point. The crevasses were one of the greatest threats and were often partially covered by a thin snow bridge – some were so huge they appeared almost bottomless.

The worst part of the ascent was Day 4, when we were constantly being pushed sideways and off bearing; at the end of a ten-hour day we had only moved a mile or so closer to the Pole. It was the first time that we started to doubt whether there was a safe navigable route up the Axel Heiberg. Since Amundsen climbed the glacier, the geography would have changed dramatically, and although there had been a navigable route in 1911, there might not necessarily be a safe one in 2011.

The view within some of the small polar community back in 2011 was that the only way anyone should attempt to climb the Axel Heiberg with pulks should be by starting your expedition at the base of the glacier, getting dropped there by plane and ensuring any loads were fairly light. Then climb the glacier and take another resupply at the top. Hauling pulks 450 miles across the Ross Ice Shelf prior to the ascent, as we had done, without a resupply, was generally regarded as just too difficult.

By this stage of the expedition, it also became clear to Henry and me that we had got our nutrition planning very wrong. The balance of carbs, fats and proteins required to meet with the physical demands of a polar expedition simply did not exist within our diet. Each of us required around 6,000 calories a day to meet the physical demands of pulling the pulk and generating enough body heat to stay warm.

In truth, neither Henry nor I had really paid that much attention to our nutritional requirements. Back then I didn't fully understand nutrition, and Henry wasn't really interested – he regarded the subject as just more laborious detail. The net effect of this oversight was that by the time we reached the foot of the glacier, Henry and I were underweight and struggling physically.

After six exhausting days we reached the polar plateau, a high central area of the continent, which encompasses the South Pole. Reaching the plateau was a relief, but that in itself presented new challenges. The temperature was now much colder, often dropping below -30°C – even lower with windchill. The winds were much stronger, and for the last 400 miles of the expedition we routinely faced a headwind, sometimes so powerful that each step became an exhausting battle. The air was also much thinner, given that we were now skiing at an altitude above 9,000 feet, and there were days when even the simplest of tasks – such as piling snow onto the valance of the tent at the end of a long day – left you breathless. Amundsen had named this area the 'Devil's Ballroom' and now I fully understood why.

The more tired I became, the more mistakes I began to make, and one in particular could have resulted in the loss of the tips of my fingers and my thumb. Prior to the start of the expedition, during one of our training sessions, Henry repeatedly warned us about the dangers of getting clothing wet, either from sweat or from spilt liquid. My near-miss with catastrophe came when I'd stopped for some food and drink. The wind was howling, and as I poured some warm water from my flask into a cup, a gust knocked my hand and the water spilled onto my glove. The water seeped through the fabric and began to freeze my thumb and the tips of my fingers almost immediately. I felt a surge of pain, almost a burning sensation in my hand, as part of it instantly froze. It was a stupid mistake with potentially serious consequences. I was forced to ski the rest of the day with my hand cupped inside my glove, while trying to hold a pole, in an attempt to rewarm my fingers and thumb. I can remember them feeling distinctly wooden.

That night, in the tent, I removed my glove to see that half of my thumb was completely white and rock hard and the tips of a couple of fingers had large blisters on them. As Henry sorted out the tent, I was distracted with my hand.

'What the bloody hell's going on in there?' Henry shouted from the outside. 'Get that cooker going. What are you faffing around for?'

'I think I may have frostbite,' I said, staring at my swollen digits.

'What?' Henry responded, and stuck his head into the tent. 'For God's sake, how did that happen?'

I explained how my flask had spilt after being caught by the wind.

'Well, that's going to cause you some problems. You're going to struggle with those fingers.'

I knew by now not to expect any words of sympathy, and laughed as I rubbed aloe vera into the damaged part of my hand before wrapping it in a bandage. I now faced the challenge of ensuring I didn't let my hand refreeze for the remainder of the expedition. Easier said than done in -30°C temperatures and still potentially three weeks from the South Pole.

For the first time I felt a sense of doubt beginning to creep into my consciousness, and I began to wonder whether I had the mental strength to complete the expedition. The prospect of day upon day of dragging my pulk across Antarctica for at least another three weeks was a desperate one. Antarctica was beginning to expose my weaknesses, just as it does everyone's, and has done since the first explorers arrived. But whilst my morale was dented, I was also determined that I would not give up – I was not going to be beaten, I told myself, and drew heavily on Stirling's ethos of self-discipline.

During our time on the Ross Ice Shelf, Henry had mostly skied with his face exposed, whereby I had religiously kept my face covered with either a buff or balaclava. All seemed fine until we got onto the polar plateau, with its much lower temperatures. I'm guessing his skin must have got damaged from all the UV light exposure, and he started to suffer with cold injuries in his cheeks and chin. At the end of one particularly cold day, he had solid areas of frozen tissue on his chin. I treated them with aloe vera gel that evening in the tent, but was woken several times in the night by Henry groaning in his sleep, clearly in immense pain. Over the following days he cursed his own poor judgement, but never once sought sympathy.

The polar plateau gave us the chance to make up some lost time. The days were long and hard but, in comparison with the earlier part of the expedition, they were relatively uneventful. It was just a case of head down and grind out the mileage. The last 300 miles was Groundhog Day on steroids – the only variable was the weather. Henry and I were both now running on empty, but we also knew that if we did everything right then we would succeed. My days took on a familiar routine: ski for an hour up front – five-minute rest, slot in behind Henry, ski for an hour, five-minute rest and repeat. The twelve hours would sometimes fly by as I listened to an audiobook or one of the 8,000 songs I had on my iPhone.

On Day 67 we were both aware that the Pole was about to come into sight. Both of us had been scanning the southern horizon for the smallest of specks. It was just after lunch that Henry spotted the station.

'There it is, Lou,' he shouted excitedly, pointing into the distance.

'Where? I can't see it.'

'There, directly ahead of us. Look – it's just a speck but there it is.'

Then, almost as if by magic, I locked on to what was little more than a black dot. I sank to my knees, offered up a prayer of thanks, and sobbed quietly inside my goggles and hood, hoping that my lack of emotional restraint would not be seen by Henry.

The Geographic South Pole is home to the Scott Amundsen National Science Foundation South Pole Station, where around 200 scientists and support staff live and work through the summer season. One of the areas being studied was neutrinos – subatomic particles, similar to an electron but without any charge, which travel through space. When they hit the Antarctic ice sheet – which is about 9,000 feet thick at the Pole – they slow down sufficiently that the scientists can study them.

It was another perfect cloudless day, and the sun shone like a golden globe in the sky. Gradually, as we skied closer, the complex began to take shape – the huge, white observatory building, then the main complex where the scientists spend their days. I could see Henry was eager to get there, but with several hours of skiing still ahead of us, I needed to stick to the routine of taking breaks for food and water, such was my exhausted state.

'Come on, let's get in there and have a hot coffee and some fresh food,' Henry said, trying to encourage me to speed up. But I was determined to savour this moment. I wanted to arrive in a reasonable state, not completely done in.

Before setting off, ALE had advised us to approach the South Pole on a particular bearing, but we couldn't remember why. When we realized we were on the opposite side of the complex, we elected to ignore ALE's advice and straight-line it in. We started to see a few people come out of the building and wave, and assumed it was a welcome party. As we got closer it became clear that they were actually quite irate. Unbeknown to us we had just skied straight through a designated clean-air sector where various scientific instruments were installed, including very sensitive seismic ground sensors, and we had triggered a bit of a reaction inside the station. We quickly apologized and Henry, working his magical charm, was able to swiftly defuse the situation, particularly when they realized we had travelled over 800 miles from the Bay of Whales.

Henry and I then skied the final few hundred yards to the actual Geographical South Pole marker, a wooden staff around four feet high. The South Pole marker has to be moved every year to represent the bottom dead centre of the planet. This is because the ice sheet is flowing towards the coastline at a rate of approximately 13 feet a year.

Henry very graciously stopped just short, allowing me to ski forward and place a hand on it first. I was almost overcome by a huge sense of relief and suddenly felt totally drained. I hugged Henry and thanked him for everything and then we each lit a celebratory cigar. A member of the US station team quickly appeared to take photographs and to ensure all was well.

'Are the Scott Team here?' Henry asked as one of the ALE staff appeared.

'No. You are the first in, congratulations. We think they are still several days out.'

'Yes, stick that in your pipe and smoke it!' Henry shouted triumphantly, and I began laughing.

'We've beaten Antarctica, Henry, we've conquered her,' I said, grinning.

'You never conquer Antarctica,' Henry said quickly. 'I remember – before my first expedition – saying how I was going to conquer the Pole during a speech I gave at a fundraiser. In the audience was the renowned polar explorer Robert Swan. After the talk, Robert took me to one side and in a very stern tone said, "My boy, do not ever be so arrogant as to say you have conquered Antarctica. If you are lucky, she will let you in to achieve your objectives, but no one will ever truly conquer her."'

That was a valuable lesson in humility that has stuck with me ever since.

Henry and I took the obligatory photos of ourselves celebrating our arrival at the Pole, and slowly I began to drink in the enormity of our achievement. It was good to win but it was better to finish.

'What happened to the Scott team?' I enquired. 'We thought they would be days ahead of us, given the terrible start we had.'

It transpired that Mark and his team had become seriously delayed after attempting a shortcut through the Shackleton Icefalls at the top of the Beardmore Glacier. It is an area of notoriously disturbed ice, riddled with crevasses, and best avoided even in good weather conditions. However, the team got caught in a whiteout and were unable either to negotiate a safe route through or to backtrack along their original line. They were trapped on a narrow ledge, unable to move, and had no choice but to pitch their tent and wait for the weather to clear, which it did after several days.

Over the next few days, while we waited for the Scott team's arrival, Henry and I enjoyed the comfort of real food, not dehydrated, and decent coffee. We slept in our tent and tried to relax and rebuild our strength as much as possible. Both of us had lost around 20 kilograms in weight – we were seriously undernourished and underweight. We looked completely out of proportion – large heads on skinny, emaciated bodies. Our clothes had become loose and I had to resort to reducing the size of my waistband on my thermal underwear, with a few stitches to stop it falling down.

Henry and I slept for twelve, sometimes fourteen hours a day over the next few days, but we quickly grew bored. There was little to do at the Pole apart from rest, read and eat – a stark contrast to the previous two months where our days had been full and demanding. Enforced idleness did not come easily to either of us.

The plan was for both of us to remain until the Scott team

arrived – they were at least nine days behind us. But the Regiment wanted their pound of flesh. As far as they were concerned, I had just been on a three-month skiing holiday and now I had to return to work. I hitched a ride on the Basler with some other expeditioners going back to Union Glacier, then on to Punta Arenas and eventually, several days later, got back to an emotional reunion with my family in the UK.

But Henry was at the Pole to greet the team. Through an incredible feat of physical endurance, they covered a huge distance in the final few days to arrive on 17 January: exactly the same day as Scott and his men one hundred years before.

After returning from the Antarctic in January 2012, I had a couple of days' leave before returning to my job in Hereford. It seemed inconceivable that just a few days earlier I had been at the bottom of the world, skiing across the snow and almost delirious with joy that weeks of pain and hunger had come to an end. Back at Hereford there was minimal interest in our achievement, not through any malice but simply because operational life was so busy. I suppose I had thought a few people might offer their congratulations and give me the odd pat on the back, but it was not to be. It's probably also true that – unless you have a real interest in polar travel and the challenges expeditions face – it is very difficult to grasp what the reality involves and what a huge achievement it was. I tried not to take the lack of interest personally.

I was delighted that Mark and the guys had finished safely and managed to get to the Pole as an intact team. There were all the usual post-exercise reports to be written up, some celebratory dinners, meetings with our sponsors, and the good news that the expedition had raised £150,000 for our chosen charity: the Royal British Legion. But gradually Antarctica slipped from all our minds as we became embroiled in the routines of work and family life once again.

9

GLUTTON FOR PUNISHMENT

Polar exploration is at once the cleanest and most isolated
way of having a bad time which has been devised.

APSLEY CHERRY-GARRARD

I was insistent that there was no way I was ever going to return
to the South Pole.

Shackleton was right when he suggested that Antarctica
exposes the very nature of a man's soul. The extremes of surviving in the most hostile place on earth does something to a
person. It changes you. It makes you aware of faults in your character, failings you didn't know existed; it exposes your every
weakness.

The simple act of getting up every morning – knowing that
you have nothing in front of you but hard, demanding, painful
work for the next fourteen or fifteen hours – is a deeply unpleasant experience. It takes you to a dark place. Antarctica has the
ability to break the soul, and mine had come close to snapping.
Such expeditions are also always selfish endeavours. Loved ones
get left behind; they become secondary figures in your life as you
obsess about equipment, nutrition, skis and training. They have
to live with the fear of not knowing whether they will see you
again, and are expected to slot straight back into your routine
when you return. My family were at least used to dealing with
this every time I went away on an operational tour, but it didn't
mean they found it easy.

I had suffered more in those nine weeks in Antarctica than I had in my entire twenty-six-year military career, including twenty years in the SAS, and I had no desire to return. It took me around four months to return to full fitness and to regain the lost weight. But the human consciousness is a great filter and, slowly, as the months passed, I forgot about the pain, the cold and the sheer exhaustion. I had the urge to go back, to challenge myself again, to learn from my mistakes and perform better – the unrelenting pursuit of excellence. But life got in the way. I wasn't a professional adventurer nor a full-time explorer. I was a member of the Regiment, and there was a global war on terror in full swing.

After my operations role finished, in mid-2012, I was offered the chance to work in a headquarters in London for twelve months. It was very interesting but sensitive work at the heart of the country's national security. When the tour in London came to an end, I was promoted to warrant officer 1st Class (WO1) in 2013 and was asked to help establish a new combat survival training centre at RAF St Mawgan in Cornwall. Up until that point, each individual service conducted their own forms of combat survival. The RAF had its own survival school at St Mawgan, where aircrew would be trained in what to do if they were shot down over enemy territory and how to survive until they could be rescued. The Royal Navy had a Sea Survival school at their HMS Sultan base in Gosport, Hampshire, and the Army had a Resistance to Interrogation unit at the Intelligence Corps HQ in Chicksands, Bedfordshire. The plan was to create a single, Tri-Service Defence, Survive, Evade, Resist and Extract Training Organisation (DSTO). I was to be the newly formed unit's first regimental sergeant major (RSM), while the commanding officer was from a naval background.

As well as unifying a tri-service training syllabus, the new school would be responsible for running the Special Forces selection Resistance to Interrogation phase – hence the need for the entire operation to have an SAS warrant officer overseeing the training. The unit has gone from strength to strength, incorporating ideas and philosophies not just from the Special Forces but also the RAF and the Royal Navy. Today the school runs one-day briefing packages for civil servants going into hostile environments such as Somalia or Libya, right through to two-week courses during which troops are on the run across Exmoor, living off the land and trying to avoid capture.

For most students, the opportunity to take part in an escape and evasion exercise was a bit of fun and a change from the daily routine of service life. But occasionally there were a few men and women who were struck by a crippling sense of anxiety. To put their minds at rest, I would recount a tale from my early military career when I had done some similar training. I explained that when I did the exercise, it started with several of my mates and I being jumped by the instructors, who were hidden behind trees. All of us were wrestled to the ground, where we were plasti-cuffed, blindfolded and told not to speak. One by one we were frogmarched into a building, strip-searched and issued with a massively heavy woollen Second World War battledress, our underwear, socks, and boots without laces.

Some of us had resorted to fairly drastic measures to make life a little more comfortable. I had folded up a £5 note into a small square, tied it with cotton, put it inside the tip of a condom and swallowed it in the hope that it would emerge several days later and I'd be able to buy some food from a local shop – at least that was the plan. The problem was that it emerged too early and I had to wash it off in a stream and swallow it again. I convinced myself it was for a worthy cause, and was worth the joint risk of falling ill or being caught.

Later that evening, still blindfolded and with our hands bound, five of us were dragged out of the building where the remaining members of the course were being held, and dumped on the floor of a Transit van before being driven off, deep into the local countryside. Any attempt to talk or to establish who I was in the van with was met with a gruff 'shut up' and a sharp kick to the ribs. About an hour into the journey, I felt a hand place something in my greatcoat pocket, followed by two taps on the shoulder – as if an instructor was saying: 'Here's something that might help later.' I hoped it was food and, feeling my morale lift, I smiled to myself in the belief that a sandwich was waiting for me to consume. None of us had really eaten properly for a few days and my stomach was constantly groaning.

For the next two hours, the van seemed to be hurtling across the entire county, swinging round the bends of small country lanes while we were being tossed around in the back, feeling increasingly nauseous. Then, as if the driver had suddenly become bored, the van screeched to a halt and we were thrown out onto the ground, completely disorientated.

An instructor removed our blindfolds, freed our hands and provided us with a scenario in which we should now assume that each of us was an escaped prisoner of war, on the run from enemy troops who were under orders to capture us. Instructions on how to avoid capture would be given to us by agents at a series of checkpoints.

'Your first agent rendezvous is in this churchyard here,' the instructor said, jabbing a fat, dirty finger at a sketch map that looked as if it had been drawn by a four-year-old with a blunt pencil.

'It's about 12 miles away and you need to get there by 3 a.m. tomorrow. You'll receive more instructions when you get there. This should help,' he added as he handed over a small compass.

His final words were: 'The hunter force is the Parachute Regiment and they won't be pulling any punches.' And with that he jumped into the van and disappeared into the night.

The five of us quickly moved off the road, into a field and crouched behind a hedge. We looked at the map, orientated ourselves, and then had a good laugh about what had happened over the last few hours. Then I remembered my sandwich.

'Hey guys. I think I've got something to eat – one of the instructors put something into my pocket,' I told them excitedly.

In the moonlight I could see their eager faces – we were all very hungry. I reached into my pocket and pulled out a flattened rat – it was roadkill. There was stunned silence.

'I think I've got something as well,' someone else said and he pulled out a porn DVD. We all had a bit of a laugh as we realized it was just the instructors winding us up.

It was quickly decided that the best way to avoid capture was to move at night and avoid all roads.

That night my team covered around nine miles and, as dawn began to break in the east, we hid ourselves in some woods on the slope of a hill overlooking a huge valley below. The valley became our general axis of advance, and made navigation a lot easier, but we also knew the hunter force would be using night-vision equipment, so we still had to take care and move tactically.

The afternoon of Day 3 found my group in another large wood overlooking a river, where – in the distance – a garden party in the grounds of a nineteenth-century manor house was under way. We all watched as families sat on picnic blankets and around tables, eating sandwiches and cakes, drinking Pimm's, wine and beer and demolishing a large hog roast. Children were running around on large manicured lawns while some of the adults played croquet. It was a scene almost from another era, and our stomachs began to rumble with envy.

Most of the local householders, farmers and landowners had been told that a military exercise was taking place and they were all politely asked not to help any of the soldiers who might appear, however hungry and bedraggled we looked. Fortunately for us and most of the other guys on the exercise, the request was roundly ignored and the civilians in the area were only too happy to help.

Given that we hadn't eaten in about four days, it was decided that one of us should venture down to the party and politely ask if we could scrounge some food. Although we had been taught how to snare rabbits, the reality was that we were always on the move. The instructors had warned us that if anyone was caught cheating, they would fail the course. In reality everyone cheated – or at least tried to. The five of us spoofed to see who the unlucky soul would be to venture down to beg for rations. Pash was the unlucky loser, and he headed down to the party with our food orders ringing in his ears. He gradually made his way into the garden and across the lawn, where the owner welcomed him with open arms.

'Come and join us, my boy,' Pash later recalled him saying. 'You must be one of those military types on exercise. We have been told not to help you, but mum's the word.' The portly land-owner tapped the side of his nose.

We watched enviously as Pash set about eating like a man pos-sessed, while at the same time stuffing sandwiches and lumps of cheese, crisps, sausage rolls and delicately made cucumber sand-wiches into his greatcoat pocket.

As he demolished plate after plate of food, the owner turned to him and said: 'Look, there are some more of your friends.' He pointed at members of the hunter force who were in a field on the other side of a hedge at the far end of the garden.

'Come and join us. You chaps, come and have some food. One of your lot is already here.'

Pash gasped in horror, hurriedly thanked his hosts for their generosity and ran across the lawn, where families were enjoying the afternoon sun, followed by six members of the hunter force in hot pursuit, ordering him to stop.

As the hysterical scene played out down in the valley below, the rest of the team became convinced that Pash was going to be caught and kicked off the course. He led the Paras on a merry dance, eventually losing them after swimming across a fast-flowing river, a risk the hunter force was not willing to take.

Pash eventually returned at around 8 p.m. that night, soaked and exhausted, just as daylight was beginning to fail.

'Where have you been?' I asked, amazed that he hadn't been captured.

'I led the hunter force away from you guys on a massive deception loop and then tracked back,' Pash responded.

'Great, now where's the food?'

Pash reached into his pocket and pulled out handfuls of soggy mush and cheese decorated with pieces of cucumber. The cocktail sausages had also disintegrated, along with the crisps, which had become a sort of salty dust. We looked at one another, the smiles slipping from our faces.

'Sorry,' Pash said, 'it was the river crossing.'

But we cared not, and stuffed the slop he held in his hands into our mouths. It was disgusting. A sort of cold, wet, sickly bready dough that tasted of river water. I gagged but just about managed to stop myself from throwing up. At least it was something, I reassured myself.

As darkness eventually entered the valley, we continued on to the next rendezvous with the agent. But this time we were ambushed by the hunter force who had been lying in wait. Once again we were blindfolded, tied up and pushed into a cattle truck, where the remainder of the course participants were already being held. It was tough, but a thoroughly enjoyable course overall and I learnt a huge amount.

In late 2014, after two years in Cornwall, I became the RSM of a Reservist unit. I was pleased to get another job as an RSM but, if I'm honest, initially I wasn't overly enthusiastic about working with the Reserves. By that time I had been a member of the regular SAS for twenty-two years, working alongside some of the most capable and professional soldiers in the world. To be told that I was now going to be running a regiment of Reservists was a step into the unknown for me.

The Regiment's headquarters were in an area of Birmingham plagued by high crime and unemployment rates. A stale air of neglect hung over the place, and I knew immediately that it would need more than a lick of paint to turn the Regiment into an efficient and effective unit. While the headquarters were in Birmingham, the squadrons were located in the northeast and northwest of England and Scotland.

I arrived around the same time as the new commanding officer, and we immediately gelled. The headquarters consisted of a small team composed of myself, the CO, an adjutant, a quartermaster and regimental quartermaster sergeant (RQMS), two permanent staff instructors, and a few support staff. During the week, the unit often seemed deserted and gloomy, and was about as far from my former role in Hereford as you could get.

At the time the Army was going through a period of tremendous change. The regular Army was being reduced in size from around 102,000 to 82,000 while, under a plan known as Operation FORTIFY, the Reserves were going to be increased from 12,000 to around 30,000 by 2020. Clearly there was tremendous pressure on every Reserve unit to boost its recruitment and retention figures.

Almost immediately the CO, me and the rest of the staff began to draw up a series of measures to help meet the aims of Op FORTIFY and raise the unit's corporate identity. The Regiment's history dates back, through various incarnations, to the Second World War, when it was known as MI-9 – Military Intelligence 9. One of its roles then had been to help downed airmen escape from occupied Europe by feeding them through the Resistance pipeline, all the way back to the UK. Those aircrew who did make it home were awarded a badge that consisted of three witches on broomsticks, signifying being spirited away in the night; we decided to reinstate that as the unit sports team badge. The base was a pretty unpleasant place to visit, and the Reserve soldiers would spend hours travelling into Birmingham for a few hours during the week or at the weekend, so we decided to upgrade the facilities to make it more appealing. The gym equipment was updated, the toilets and showers refurbished. All of the training that the unit was delivering was overhauled, with the aim of raising standards across the board.

Although my workload was initially quite heavy, once the plan for the Regiment had been developed, there was relatively little for me to do. Most of the training was conducted by the Permanent Staff Instructors and, apart from dealing with the odd disciplinary matter, I shadowed the CO as he visited the squadrons and met the troops.

The soldiers were an eclectic bunch. There were doctors, lawyers, bricklayers, architects and artists. There were also former members of the Army and guys who were unemployed and relied on the Reserves to help make ends meet. Some of the guys were fully committed, religiously attending weekend training and drill nights every week, while others seemed to disappear for months on end. But most importantly they were civilians, not regular soldiers, and they required a more nuanced form of leadership.

It was at that stage that the CO asked me to have a look at the adventure training side of regimental life.

'How about a major flagship expedition which would raise the Regiment's profile and encourage recruitment. Something really big,' I said, almost without thinking.

'Brilliant,' the CO said, clearly enthusiastic. 'Any ideas?'

'I've previously skied to the South Pole as part of a military expedition. How about I lead a team of Army Reservists to the South Pole again?'

'Fantastic. We could get loads of people involved with the planning and the training. It would really raise morale and help break the mould of normal training, and it could really put the Reserves on the map.'

'The main challenge would be the financing, but I think there are various bodies within the Army and the Reserves that will help with that. The key is team selection. Plus, anyone volunteering would have to contribute a lot of time and some of their own cash.'

I was shooting from the hip and hadn't really given any thought to what I was saying.

'This is brilliant, Lou,' the CO enthused, 'but are you sure you can do this? Don't tell me this if you don't think you can deliver.'

But I was adamant: 'It will be a huge project, but it is achievable.'

As soon as I made the decision, I felt incredibly energized and excited. Over the weeks that followed, the momentum surrounding the expedition grew and took on a life of its own. It soon became bigger than *Ben Hur*, and then I had a bit of a wobble. I realized that all of the planning would rest on my shoulders, and I also began asking myself tough questions about whether I had the capability to lead a team to the South Pole. The last expedition had been tough enough, without all the added pressure of leading a group of Reservists across 730 miles of the most hostile terrain on earth. But my doubts were soon expunged as I began to put the regimental motto of 'Who Dares Wins' into practice. It was now up to me to pick and train the right team and make sure I had a workable plan.

The CO and I decided that we would open it up to all members of the Reserves and attached personnel. Once the word began to spread, there was a real buzz around the units. Lots of people began asking me questions and told me that they were intending to apply for a place. It was at that stage in February 2015 that I realized the time had come to formalize the planning.

I organized a briefing day in Birmingham, where more than

sixty potential volunteers from various units came to listen to a presentation by me about the reality of taking part in a polar expedition. I decided that – rather than give a hard sell – I would do the opposite, and attempt to deter anyone who harboured the slightest doubt.

'Thank you very much for coming along today. I know you are all very interested in taking part in the expedition, and will probably see this as a once-in-a-lifetime trip to a continent like no other on earth. The concept is to ski from the inner edge of the Ronne Ice Shelf to the geographic South Pole – a distance of 730 miles. The plan will be to leave the UK towards the end of October 2016, and complete the journey some fifty days later. One of the aims of the expedition will be to raise at least £50,000 for ABF The Soldiers' Charity, so this will be a serious undertaking. But first of all, I need to lay out some hard, basic facts.'

It was at this point that I saw a few smiles drop from faces.

'I am looking for just five people – that's all. You will have to contribute £5,000 of your own money and will have to take at least three months' leave from your places of work. It is a massive commitment. There will also be a two-week training session in Norway for those shortlisted to form the final team.

'I really want to impress upon you how hard this expedition will be. It will be unsupported: this means you will carry everything you need, about 120 kilograms of kit, food, fuel and clothing, in a pulk – a type of sledge that you will drag across the ice for over 730 miles from the coastline of Antarctica to the South Pole. The temperature will drop at times to -40°C and you will face winds of fifty miles an hour. You will be at risk from frostbite, exposure and snow blindness, and under constant threat from deep crevasses. As well as skiing across more than 700 miles of ice, the team will also have to climb from sea level to more than 9,000 feet in height. The expedition will be the hardest thing you will ever do, and I can tell you from personal experience that it will be far harder than you can possibly imagine.

'I took part in an expedition to the South Pole with five other guys in 2011. We were all very fit and had served in just about every type of operational theatre going. One member of the team lasted two days. The rest of us were pushed to our mental and physical limits. When I finished the expedition, I had lost 20 kilograms in weight and I vowed never to return.

'Now, imagine the hardest physical thing you have ever done in your life. For some of you that will be Reserve selection. Now imagine doing that for twelve hours a day, every day without exception for eight weeks, in the most hostile environment on earth. That is the sort of challenge you will be facing.'

I stopped talking and looked around the room. At least 30 per cent of those attending the brief had already changed their minds.

'The expedition will not be for everyone. There are only five places, but everyone in this room can play a part if they want to. There is a lot of planning and fundraising to do, and your contribution will be vital.'

At the end of the presentation I gave everyone still interested a form to complete, asking for their personal details, what relevant experience they had and why they wanted to come on the expedition. I also explained that there would be a series of assessment weekends in different parts of the country where I would whittle down numbers and begin to produce a shortlist.

'Just one last point,' I told the audience. 'The people I am looking for will have to work as a team – that is the only way the expedition will achieve its goal. You will have to be fit but you will also need to possess the type of personality that can deal with problems when everything goes south. So, I will be looking for that as well. Good luck.'

Over the next few weeks, the expedition planning took on a momentum all of its own. It became an all-consuming undertaking. I started planning the selection weekends in detail and began the lengthy process of fundraising. The expedition was going to cost around £400,000, and I soon began submitting applications for various grants to numerous different bodies within the armed forces which specialized in funding Reservist expeditions.

The Army has a wonderful self-help attitude towards adventure training, and because my expedition involved Reservists, I found that many people within the organization were only too keen to help.

The more I researched the subject, the more I discovered that there were plenty of pots of funding around. I applied for grants to organizations such as the Ulysses Trust, which primarily supports Army Reserve expeditions. I was also very generously supported by Lieutenant-General Ty Urch, who went on to become the head of the Army's Home Command. He also came to the rescue during a minor funding emergency. I had pretty much secured all of the necessary finances to pay the invoice from ALE, which was in US dollars. However, a month before I was due to pay the main bill, the uncertainty surrounding Britain's future exit from the European Union saw the pound plummeting against the dollar, adding £30,000 to the bill. The expedition really was hanging in the balance as I desperately scrabbled around trying to secure the extra money with days to go before the payment deadline.

As with all previous expeditions, going back over a century to the days of Scott and Shackleton, fundraising was always a tough challenge, and so I began the equally lengthy process of seeking patrons who would back the expedition. The bigger the name, the greater the chance of raising the necessary funds. So it was time to think big. My first port of call was the Royal Family, but sadly that didn't work out. Instead I contacted Sir Ranulph Fiennes, who said he would be delighted to help out.

Meanwhile, the selection process began. Over four separate

weekends, I planned to whittle down the fifty or so Reservists still interested, into a workable group of between eight and ten from which the final five would be chosen.

The format of the selection process was fairly brutal but fair. The first weekend selection process took place at the Warcop Training Area on the edge of the North Pennines. The main part of the day consisted of dragging a Land Rover tyre on a harness around a two-mile course for seven hours. One of those who arrived for the selection weekend was Captain Eva Howard, a doctor. Eva was one of the most effervescent and positive soldiers I had ever come across; she was also super-fit and loved a challenge. I knew that she would make a brilliant asset to any expedition, but I feared that her diminutive frame – she was around five foot five inches tall and weighed less than 50 kilograms – would count against her. Within an hour of the start of the tyre-hauling, people began to drop out. I noticed immediately that Eva began to fall back – she just didn't have the physical bulk to pull the weight. By the late afternoon most of the candidates had finished, but Eva was still out on the course, refusing to give in. She eventually finished the course way behind the last of the men. After a shower and dinner, more tests and interviews took place late into the evening, followed by further physical tests the following day.

The following week, I returned to headquarters and gave the CO a debrief on the weekend.

'I really wanted Eva to make it onto the team,' I told the CO. 'She would be great for morale, it would be good for diversity, and she is a very tough soldier. But my two main rules were that candidates had to complete the course in a set time and that everyone carried the same weight. If she was in the team, she wouldn't be able to keep up with the pace set by me and we couldn't afford to go any slower. I've no doubt she could get to the South Pole if time wasn't an option – but it is.'

The CO nodded and I could see that he was just as

disappointed as me. She was one of those people who really lit up the room with her presence and smile.

I continued, 'I could reduce her weight, but I don't think she would agree, and neither would it be fair on the rest of the expedition members.'

Later that day I broke the news to Eva by phone. I could tell that she was disappointed, but in her heart she said she knew that she had probably not made the grade.

By the end of the third weekend, I had covered polar navigation, cold-weather injuries and basic expedition skills. From the remaining twenty or so guys, I selected a shortlist of eight; coincidentally three of them were junior doctors and one was a paramedic. They were a mixed bunch from different backgrounds, but they all seemed to gel and accept each other for what they were. It was precisely the attitude I was after.

The final eight included Ollie, a super-fit ultra-runner and a junior doctor, who was probably the strongest of all the candidates, although I didn't tell him that at the time. He seemed to possess all of the right qualities – humour, physical resilience, good judgement and intelligence, and he was a good team player.

Alex, another junior doctor and the son of an MP, had also fared well during the selection. He was very bright and affable, but he had very decided opinions as well. Our third junior doctor, James, was a very bright, well-balanced soldier who was equally at home being either a leader or a follower – two qualities that were both necessary. The physically most able amongst us all was Chris – a paramedic who weighed around 100 kilograms of pure muscle. Chris was a super-fit boxer from Yorkshire, and although physically imposing he was the quintessential gentle giant. Next came Alun, a web designer from Anglesey. Alun was a free diver and could hold his breath for up to seven minutes. He was extremely intelligent and held a master's degree in quantum physics. In many respects he was the odd one out within the team. He was a quiet introvert who rarely voiced an opinion, but he was a very likeable guy and was keen to offer his graphic designing skills in every way possible to help promote the expedition.

Another member of the shortlist who had also done very well on the selection phase was Steve, a proud Yorkshireman and a fireman. Steve was very different to the rest of the team. He was loud and brash and acted as a great foil to the highly educated and charming doctors. There was also a filmmaker called Matt and lastly Ian, a long-serving sergeant major who was the oldest member of the team.

I explained to them all that the five places were still up for grabs and that anyone who wasn't selected would become a reserve, so the entire eight would undertake all of the training including two training trips – one in Norway, which would be similar to the training I undertook with Henry in 2011, and another in Iceland, where we would practise crevasse rescues and give our kit a final test run before shipping it to Punta Arenas in Chile.

Returning to the never-ending process of fundraising and identifying possible patrons, I decided that Julian Brazier, who was the Minister for Reserves, was an ideal candidate as a patron, a position that he enthusiastically accepted. The third patron was General Sir Cedric Delves, a former director of Special Forces and president of the SAS Association.

With the patrons now formalized, Alun began working on a website and logo, and we decided on a name for the expedition: it would be known as the South Pole Expedition Army Reserves 2017 – or SPEAR17 for short. We wanted SPEAR16, as that was when the expedition was due to start, but when we checked the availability of the domain name, it had already been taken by some ancient battle re-enactment group. So, we went for the next best thing and the year that we would complete the trip.

Just as I was getting into fundraising, Henry Worsley was on the verge of returning to Antarctica, on his Shackleton Solo expedition, in a bid to become the first person to undertake a solo, unsupported traverse of the continent. He had spent the previous two years in Washington working as a liaison officer, but his Army career had just come to an end and he was about to embark on a truly bold and ambitious final polar journey. His preparation had gone well, and some wealthy backers had stumped up the cash he needed for the trip. Henry was incredibly excited about the journey, and I truly believed that if there was one person on the planet capable of skiing unsupported across Antarctica, it was him.

Our busy lives meant that we hadn't seen much of each other in the years since the Scott–Amundsen Race, but we kept in regular contact by email. Henry left for Punta Arenas in October 2015, and my last words to him were wishing him the best of luck on an incredible undertaking and, using his own advice, to 'keep putting one ski in front of the other.'

The actual expedition began on 13 November, and he kept the world informed of his progress through an audio diary that appeared on his Shackleton Solo website every morning. No matter what the conditions, no matter how tired he was, Henry would record a three- to four-minute clip via his satphone. His diary became compulsive listening for everyone vying for a place on the SPEAR expedition. I would wake up around 6 a.m., grab my iPad and listen to his latest blog while lying in bed. I could almost picture myself sitting in the same tent as Henry, going through our evening routine, perhaps drinking some hot chocolate and sharing a few jokes. I was utterly convinced back then that Henry would succeed. I don't think the thought of failure ever entered my mind.

In early January 2016, the nine members of SPEAR17, plus two support staff, arrived at the base in Norway. I hadn't let on that this was where the final team selection would take place, although I think most of them had a good idea who were the front runners.

One of the greatest challenges for the leader of any expedition, especially those to the more extreme corners of the globe, is team selection. I wanted leaders and followers and people who could do both. Physically, and almost more important mentally, they needed to be able to cope with undertaking an enormously demanding challenge in one of the most extreme environments on earth, every day for nine relentless weeks.

Just like Shackleton and the other great polar explorers, I was gifted with a team full of riches, and trying to choose the best five men from eight was far harder than I had anticipated.

The first two days of the Norway trip were spent getting to grips with our new equipment. The pulks were packed, harnesses adjusted, our cookers were tested, taken apart, serviced and put back together again. There were lessons on first aid and pitching the tents, and we spent a lot of time getting to grips with cross-country skiing. Given that we were going to be self-sufficient and unsupported, it was also crucial that everyone knew how to care, maintain and repair all of their key pieces of equipment.

The atmosphere was great and morale was super-high, but the weather was deteriorating and the temperature had dropped to around -25°C – extremely cold, even for Norway, in January. In fact, it was so cold that the RSM of the Norwegian base, who was acting as our host, suggested we postpone our expedition on Lake Femund for a couple of days. We didn't have the luxury of time, though. We were heading off to the Antarctic, so to delay the expedition because it was too cold wasn't really on the cards.

The lake had been frozen for several months and the ice was around 13 feet thick. It was covered with several inches of snow and flanked by jagged snow-capped mountains. A more beautiful Norwegian winter scene was difficult to imagine. But it was painfully cold, so much so that it actually hurt to breathe, and the threat of frostbite and hypothermia was very real. But the weather did not deflect from the overwhelming excitement felt by every member of the group as we arrived at the lakeside.

Everyone was buzzing. I could see that they all felt like real polar explorers for the first time.

Once we had unloaded our kit from the minibus, I quickly worked out a bearing, confirmed everyone was ready, and we set off north right down the centre of the lake.

By 11.00 p.m., everyone was in their tents, having cooked their rations, made hot drinks and chatted about the day's events. I was sharing my tent with Ollie on the first night, although my tent-mate would change every evening so that I could get a bit of one-to-one time with them all. What I hadn't told the rest of the guys was how noisy the ice can be. It creaks, snaps, cracks and booms – almost like gunshots – as it constantly contracts and expands, and for the uninitiated the sound can be extremely disconcerting. I thought it would be funny to see how the team would respond. I didn't have to wait long. As the conversations died away, there was a short period of silence before the first bang split the still winter night.

'What the hell was that?' someone said in another tent.

'The bloody ice is cracking.'

'It can't be – Lou said it was 13 feet thick this morning.'

Head torches suddenly came on as some of the guys began to emerge from their tents to check the ice. The noise seemed to settle, and the concerned voices fell silent before there was another ear-splitting snap.

'Jesus Christ – this isn't safe. We should wake Lou,' someone else said.

I could hear the entire conversation, but decided to act as if I was still asleep. Ollie was already sitting upright and whispering to the other tent.

'I don't want to wake Lou – he was saying that it was really important to get a good night's sleep.'

'I know,' said a voice, 'but this is really bad.'

I could hear the zipper on my tent being undone.

'We need Lou. This is serious. We have to get off the ice now,' someone said quietly.

I was doing my best to remain still and not to laugh.

'I know. But look at him,' Ollie said. 'He's completely out. He's oblivious to it all. I think we should leave him. Think about it. Lou's not going to take us on a lake that isn't safe.'

They withdrew from the tent and I eventually got some sleep. The team had breakfast inside their tents, packed away their kit into the pulks and gathered around for a quick brief on our route for the day.

'Bloody hell, Lou. Did you hear all that cracking last night?' Alex asked, looking tired and bleary-eyed. 'I couldn't sleep at all. Every time there was a crack, I thought the tent was going to fall through the ice.'

'Yeah,' added James. 'I have to confess I was a bit nervous.'

It transpired that they had hardly slept at all and were all extremely tired.

I could no longer contain myself.

'Guys,' I said laughing. 'The ice is 13 feet thick. I told you yesterday. You are not going through it. You could drive a truck across this ice and it wouldn't break.'

Norway in winter is a challenging environment, but very different from Antarctica, and one of the greatest admin problems we faced was hoar frost, which forms on the inside of the tent. It is basically frozen condensation; minute shards of frosted water would shower down on top of you whenever the side of the tents were hit by a gust of wind, or if someone inside disturbed it. Falling frost would wake you at night, settle on your sleeping bag and melt, which meant that your sleeping bag became wet and almost impossible to dry. Items of clothing – such as gloves and socks – also got wet, and at temperatures of -30°C that can be a huge problem, as Ian discovered when he developed frostbite in the end of his fingers. It was only minor but it was a concern and, although I didn't mention it at the time, it effectively ruled him out of the Antarctica expedition because he would have been too susceptible to getting frostbite again.

The conditions were compounded by the darkness, which made everything more challenging, and I felt that some members of the team were questioning whether they could survive two months in Antarctica in similar conditions. I reassured them that firstly there would be effectively twenty-four hours of daylight, and secondly and probably most importantly the atmosphere would be significantly drier. There would be no hoar frost, so that items of clothing would dry out in the tent overnight. The cold was also very intense in Norway, as we were effectively camping on a giant ice cube, with the cold permeating upwards from the 13 feet of ice below us. Snow is a great insulator, but here there was only a very thin covering of it on top of the ice.

By Day 3 of the expedition, the entire team appeared to be getting to grips with pulk-hauling and skiing. When we were moving we were relatively warm, and most of the group seemed pretty content. We skied for an hour in single file, stopped for five minutes, swapped over lead man and continued. The day was going smoothly until it was Steve's turn to be lead man when, after around twenty minutes, it became clear that he was slowing down. I broke out of the file and skied up to him.

'You OK, Steve?' I asked, concerned that someone as fit and as strong as him was now obviously in trouble.

'Really struggling, Lou. I can't seem to get any speed. I'm working 110 per cent here but I'm really hanging out, I'm really short of breath.'

Steve was in a bad way. He was pale and sweating and working way harder than normal, but was going at about half the speed of everyone else. He was also very anxious, no doubt worrying what impact his physical state would have on my decision-making.

'OK, Steve, don't worry about it,' I said reassuringly. 'Let's just stop and have a rest for five minutes. We all have bad days – you're probably dehydrated or something.'

But I knew his condition was potentially much more serious. He was an excellent soldier; it was difficult to believe he would make a fundamental mistake such as not eating or drinking enough.

I quickly skied back down the file and asked James, who I had nominated as the lead medic, to check Steve out.

Steve was now sitting on his pulk, skis removed, as James began examining him, listening to his breathing and heart rate.

James returned after around ten minutes and I could tell by the look on his face that it wasn't good news.

'He needs to get to a hospital pretty soon. It's not exactly an emergency, but he should get off the lake today and get to hospital.'

As we headed to the shoreline, I contacted the support team back at the base. While we waited for them to arrive, I tried to reassure Steve that this little blip didn't necessarily mean the end of his chances of gaining a place on the expedition.

'Let's just see what the doctors at the hospital say before we make any decisions. They might say you are fit enough to return tomorrow or the following day, but the priority right now is your safety.'

Steve nodded enthusiastically. 'Cheers, Lou,' he added. 'That means a lot.'

An hour later the support team arrived and Steve was heading off to hospital. Later that night I received news that Steve was medically unfit to continue.

Over the next couple of days as the expedition progressed, I began to mentally weigh up the relative strengths and weaknesses of the remaining group. I definitely wanted Ollie and Chris. Ollie would be my second in command and had the capability to step into my shoes and take over if anything happened to me. Chris was just a phenomenal individual. He was incredibly strong, could haul three pulks on his own if he wanted to, and was called the Panzer Snow Plough by the rest of the guys.

And, just as importantly, he was great for morale. I was also fairly convinced about James as well, and particularly impressed with his swift and accurate diagnosis of Steve, which no doubt saved his life. The hardest part was picking the two spare places from Ian, Alun, Matt and Alex. There wasn't a huge difference between any of them physically – they all had their relative strengths and weaknesses, as we all did; it was more about moulding the team to get the best fit.

Alex was very bright and fit, and had smashed every task and challenge he faced. He was also opinionated and could potentially be disruptive, but it was important sometimes to have a dissenting voice, someone who will challenge the perceived wisdom and offer up an alternative plan – it was that sort of approach which had made the SAS so successful. Alun was the complete opposite. He rarely voiced his opinion on anything. He was quiet and slightly introverted, but also very bright. A great problem-solver and a fantastic team asset, while Matt was somewhere in between.

A couple of days later we were back at the base, expedition over. We sorted our kit out on the first night back in camp, and the following morning I held one-to-one interviews with the team members to debrief them on their performance. I was pretty sure about who would make the final selection, but the interviews were probably the final opportunity for anyone to potentially change my mind one way or the other.

After I had spoken individually to everyone, I gathered them in a small room.

'Before announcing the team, I want everyone to know that this was always going to be a difficult choice. All of you have the ability to take part in a South Pole expedition, and I know that you have put everything into this. So, thank you for that. Now to the team. Those coming with me to the Pole are: Ollie, James, Chris, Alex and Alun. Matt, you will be first reserve and Ian, because of your frostbite injuries, you will be the second reserve.'

The emotions of the team within the room couldn't have been more different. Ian and Matt were completely crestfallen, while the other guys were almost doing somersaults. To their great credit, Ian and Matt accepted the situation, which made it much easier for me. They also agreed to carry on with the training sessions in case one of the team members dropped out. With the final team now selected, returning to Antarctica seemed one step closer.

10

A TRAGIC LOSS

We had seen God in His splendors,
heard the text that Nature renders.
We had reached the naked soul of man.

ERNEST SHACKLETON

Henry was struggling. I had first noticed a few small signs as I listened to his audio diary during the Norway training camp with SPEAR17, but by the time the team had returned home in mid-January I was in no doubt. He was a man who didn't do self-pity. It just wasn't part of his make-up. But for the first time in all the years I had known him, he seemed to be preoccupied by some of the difficulties that come with polar travel. Earlier in the expedition, when he had been through some really tough days, he had brushed over the hardship and difficulties and focused on the positives, as he had always done. But now, something had changed.

By 20 January 2016, Henry had covered more than 900 miles and was very much on the last leg of the expedition. He still had a fair distance to go, but I sensed that he was doubting himself. His voice lacked the normal strength I had come to associate with him. During the Scott–Amundsen expedition, I had never once heard him complain, even when I knew he was suffering. He just thought pain and hardship were part of the deal, so why complain about it? He was an immensely strong individual, both physically and mentally. There were days when I was struggling and he would carry some of my cooker fuel to help out, even though he was seven years older than me.

The following morning I listened to his audio diary, and was again struck by Henry's attitude. I found myself saying: 'Just get on with it, H, you're almost there.' It was more of a passing thought than a detailed analysis. It then dawned on me that Henry was running on empty.

On the morning of 22 January, I thought to myself that Henry might not make this. I wished I could pick up the phone and give him some encouragement, but I also knew that this wasn't just a matter of will or being too tired. Henry was on the last leg of the expedition. He would have expected to be close to mental and physical exhaustion, and his preparation would have taken that into account. The key, I wanted to tell him, was get to the top of the Shackleton Glacier; from there it's a downhill run. But he knew this.

It has been reported that Henry only had 30 miles left to cover. That was not the case. He was approximately 30 miles from the top of the Shackleton Glacier, but 120 miles from his planned finish point on the Ross Ice Shelf. After getting to the top, he would then have to descend 90 miles down the glacier, a route that had only been done twice before. But he would be going downhill, the altitude would be lower, so it would have been easier to breathe. Then, once he was on the glacier, he would have been skiing on hard ice as well as being shielded from the winds of the polar plateau, and life would have been a whole lot more pleasant.

On 23 January, I woke again and listened to Henry announce that he had been forced to halt the expedition. I could tell by his voice that he was in a seriously weakened state, but I simply assumed that it was nothing that a few home-cooked meals and a few days' rest couldn't repair.

In his final message of the expedition, Henry said: 'I set out on this journey to attempt the first solo unsupported crossing of the

Antarctic landmass, a feat of endurance never before achieved. But more importantly, to raise support for The Endeavour Fund, to assist wounded soldiers in their rehabilitation. Having been a career soldier for thirty-six years and recently retired, it has been a way of giving back to those far less fortunate than me.

'The seventy-one days alone on the Antarctic with over 900 statute miles covered and a gradual grinding down of my physical endurance finally took its toll today, and it is with sadness that I report it is journey's end – so close to my goal.

'When my hero Ernest Shackleton stood 97 miles from the South Pole on the morning of 9 January 1909, he said he'd shot his bolt. Well, today I have to inform you with some sadness that I too have shot my bolt. My journey is at an end. I have run out of time, physical endurance and the sheer ability to slide one ski in front of another to cover the distance required to achieve my goal.'

I could tell that he was heartbroken, and I was absolutely gutted for him. I knew how much the expedition had meant and that he would feel as though he had failed and let people down. In reality, what he had achieved was a remarkable feat of human physical and mental endurance.

After informing the ALE team at Union Glacier that he was aborting the expedition, Henry packed up his equipment and waited for the aircraft to arrive. He managed to climb on board unaided and a cursory medical examination found that he was exhausted and malnourished but other than that he seemed to be relatively fit given the circumstances. He even posted a photograph of himself drinking a cup of tea on the way back to Union Glacier. At that stage, I was just pleased that Henry was safe and was going to get some proper medical treatment.

The following day I was returning from a motocross racing weekend with my son Luke in our camper van. It was late in the afternoon and we were on the M50 motorway around thirty minutes from home when my mobile phone rang. I could see it was a number I didn't recognize, so I decided to stop and answer.

'Hi, Lou, this is Brigadier Tom Copinger-Symes.'

Brigadier Copinger-Symes was then commander of the 1st Intelligence, Surveillance and Reconnaissance Brigade, and as we were part of that organization, my immediate thought was that there had been some sort of issue involving a member of the unit.

'Hi, boss. What can I do for you?'

'Lou, I have some bad news. I know you were a friend of Henry Worsley's and I wanted you to find out before hearing this on the news.'

In that split second before being warned of bad news and actually receiving it, I hoped that he was going to say that Henry had been taken ill or had been injured, but I somehow knew that wasn't the case.

'I have just been notified Henry died during emergency surgery. He was taken off the ice and seemed to be OK, but by the time he got back to Union Glacier his condition deteriorated quickly and he was immediately flown to Punta Arenas. I'm so sorry.'

I was lost for words. Luke heard the entire conversation and he was equally upset and shocked. Having served in the SAS for twenty-four years, I wasn't unfamiliar with sudden death, but when troops deploy on operation you are aware of the risks, knowing that people can and do get killed and injured. Henry's death was something else.

'Thank you for letting me know, sir,' I managed to say before ending the call.

I just sat there in the lay-by for ten minutes or so, trying to process what I had been told. I simply couldn't believe what I had heard. People don't die in Antarctica in the modern era, I kept saying to myself. Sure, it's a dangerous place, things can go wrong, you can fall down a crevasse, but no one had actually died on an expedition in years.

I drove home in stunned silence and broke the news to Lucy,

who was devastated; later I informed the rest of the SPEAR team that Henry had passed away. Over the next few days I learnt a little more about what had happened to him. Ultimately, he had died of multiple organ failure during emergency surgery for bacterial peritonitis. It was terrible bad luck. No doubt the physical strain of the journey, exhaustion, and the considerable weight loss were contributing factors. He had pushed himself to the absolute limit in pursuit of his goal. On the day that his death was announced, there was something of a media frenzy, but I was reluctant to say too much to the press, particularly before running it past his wife, Joanna, who had flown out to Chile to meet up with Henry.

In the following months, Henry was posthumously awarded the prestigious Polar Medal, which Joanna received on his behalf at the Palace. He had raised over £500,000 for his chosen charity, the Endeavour Fund. These two small consolations during a tragic episode would have meant the world to him.

Over the next few weeks, as events began to settle down, there was a growing feeling amongst the SPEAR team that his legacy should be honoured in some way. A WhatsApp group had been created to which all the team plus the two reserves contributed thoughts and ideas, mainly about the forthcoming expedition.

The original SPEAR aim was to ski 730 miles unsupported to the South Pole and finish there – and for a group of relative novices that was a huge undertaking – but it had been done before by other teams. In February 2016, I started to think that maybe we could have a go at the second leg. I knew the team I had were super-capable. They were all very fit and highly motivated, and I thought to myself these guys could do something really special. So I proposed a plan of starting at Hercules Inlet – rather than Henry's starting point near Berkner Island, because our route was cheaper – then skiing to the South Pole unsupported and continuing onwards on Henry's precise route over the Titan

Dome, down the other side to the top of the Shackleton Glacier and down onto the Ross Ice Shelf to the finish point. It would change from being a 730-mile journey to one of over 1,100 miles, a complete traverse of the Antarctic landmass.

I put my thoughts on the WhatsApp group and simply said: 'What do you think, guys?'

As far as I was concerned, any change to the expedition aims had to be a group decision – it wasn't something I was going to impose upon them. Being the super-proactive team that they were, though, there was never a moment's hesitation. The messages coming back were:

'Absolutely, let's do it.'

'Yeah, that is the right thing to do.'

'It will be an amazing expedition, we can do this.'

I explained that potentially we would go from being a group of guys who had skied to the South Pole, to becoming a British team to ski right across Antarctica.

'More people have walked on the moon than have traversed Antarctica. Twice as many, in fact. Only six people have ever traversed before,' I told them, 'but twelve people have walked on the moon.'

I told Joanna what we were planning, but I also said that we would only undertake the expedition, which we wanted to dedicate to Henry's memory, if she was happy. Joanna very graciously gave us her blessing and also agreed, later on, to become a patron for the trip. This meant a huge amount to the team.

The options were firstly to ski unsupported to the South Pole, take a resupply of thirty days' food and fuel and then continue on. This would allow us to have some fresh food in the ALE camp at the Pole, and maybe rest for a day or two to really enjoy the whole South Pole experience. Or we could attempt the entire crossing unsupported. This would mean carrying food and fuel for at least eighty days, which would increase the weight of the pulks to over 160 kilograms.

The team was split, with James, Ollie, Chris and Alex up for attempting the traverse unsupported. The possibility of crossing Antarctica unsupported was very attractive. But I was erring on the side of caution, warning that we didn't want to bite off more than we could chew. Alun was also with me on this.

'Right, guys, what is it that we want to achieve here?' I asked them. 'Do we want to get into the history books for an unsupported crossing, or do we want to give ourselves the best chance of reaching his final campsite and finishing the journey off for him?'

The group response was that the expedition should aim to honour Henry, and the best chance of completing that was the option which allowed for a break at the Pole. In terms of preparation, the increased mileage didn't make that much difference, but changing the finish point did have a dramatic effect on the cost, which increased by another £150,000. This was primarily due to the fact that we would end up at the Ross Ice Shelf and have to take a specially arranged flight from there, rather than jumping on board a routine flight from the South Pole at journey's end.

There were some doubters in the polar world who told me that they thought we were pushing our luck, and were questioning what I was doing leading a group of novices on a full traverse.

All of the people who had traversed Antarctica before were highly experienced professional explorers/adventurers who had thousands of miles of polar travel under their belts. I think there was a bit of jealousy about what we were trying to achieve, as well as some doubt as to whether I was competent enough to lead a group, given that I had only been to the Pole once before. But, if anything, that sort of negativity just spurred me and the rest of the team on. Besides, I had my military experience to fall back on. I was a warrant officer in the SAS and was used to leading people in exceptionally challenging and dangerous environments. I knew that I simply had to apply the same leadership skills as I would have done for a military operation.

By October 2016, the expedition planning was very advanced. The SPEAR expedition had been officially launched some months earlier at the Palace of Westminster – the Houses of Parliament – which had been kindly organized by ABF The Soldiers' Charity and Julian Brazier.

I had hoped that the Army would be able to help us with the majority of our equipment for the expedition. The Army Adventurous Training Group has a vast warehouse at Bicester containing every conceivable piece of adventure training equipment from sea kayaks to ski touring gear. After visiting the warehouse in early 2016, it became apparent that, while much of their gear was very good, it wasn't the gold-standard, niche equipment necessary to undertake a traverse of Antarctica. I was keen to avoid the issues my 'bargain boots' had created on the previous expedition. But just as I began to fear the worst, a retired major called Graham Cook, who ran the equipment side of Army Adventure Training, said that he was willing to source most of what we needed from his own budget, on the proviso that it was returned to the Army when the expedition was completed.

Graham asked me for a wish list of everything I needed to kit out the expedition, and he was as good as his word. The total bill was probably close to £30,000, but we had the skis, pulks, boots, clothing, sleeping bags and tents essential for the expedition. It was exactly the sort of help we needed, and meant that the rest of the sponsorship money could be put towards the actual cost of the expedition.

In early September, I was privileged to be asked to attend and speak at a small private ceremony to scatter some of Henry's ashes on a hilltop in Herefordshire. Looking out over the stunning view, we toasted him with cigars and a Martini. It was a fitting send-off to a great man and only served to strengthen my resolve for the upcoming expedition.

With the fundraising sorted and the patrons all on board, there was just one final opportunity to test all of our expedition equipment and practise crevasse rescue in an environment that was similar to Antarctica. The location I'd chosen was the vast Langjökull Glacier in Iceland. By now I had told Ian and Matt that in all likelihood they would not be taking part in the expedition, and as such were not required to attend the final training session. I know they were both disappointed, but I also like to think they had benefited from the whole training and selection process. And so, with just a few weeks to spare before the start of the expedition, it was just the six-man team who flew to Reykjavik on an EasyJet flight, together with our pulks and all of our equipment. I had arranged for a highly experienced Icelandic guide called Addi to pick us up from the airport and drive us straight onto the glacier.

Addi arrived in a massive 4x4-style monster truck, specially customized to traverse the glacial landscape. He greeted us like old friends as we climbed aboard. All of our kit went into an attached trailer, and off we set for the three-hour drive to the glacier, a 30-mile by 12-mile mass of snow and ice, which rises up to around 4,700 feet above sea level.

'Lou, the weather is not great,' Addi said in near-perfect English as we headed northeast across the almost lunar-like landscape that surrounds the busy airport. 'There are three very bad weather fronts coming in back to back. So, I am going to drive you to a hotel for a couple of days and then you can reassess the situation. It is too dangerous to be up on the glacier – even for me.'

'It's OK, Addi,' I said, a little too dismissively. 'We are training for Antarctica. A bit of bad weather will be a good thing. We have top-end equipment. We'll be fine.'

Addi looked me square in the eyes and said, 'Lou, you need to go to Antarctica to train for Iceland!'

Normally I would always take the advice of a local like Addi, and over the last couple of days I had been monitoring the weather on various phone apps, so I was well aware that it was at best going to be challenging. But in this case, I thought it was worth the risk, especially given the relatively short amount of time we had.

'No, Lou. I'm serious,' Addi persisted. 'I haven't seen anything like this for many years. This isn't just bad weather. This is a serious storm and it's heading right for the glacier. The wind speeds will top a hundred miles per hour and it will be hitting the glacier tonight.'

The rest of the team were listening attentively. 'What do you think, guys?' I asked. The decision was unanimous. We were heading onto the ice.

'If we can't handle the weather here, what chance have we in Antarctica?' Alex said convincingly.

'I agree,' added Chris. 'Let's just grit it out. It will be a good training session.'

'Guys, this is a really bad idea,' Addi repeated earnestly as we began to make the steady climb up towards the glacier, but eventually he was forced to relent, convinced no doubt that we were crazy Englishmen who had no idea what we were letting ourselves in for.

After a long drive up a winding gravel track, we arrived at the drop-off point near a huge ice cave that had been burrowed deep into the glacier using technology from the Channel Tunnel. It was an impressive underground complex, consisting of a tunnel with different rooms, and even an ice chapel where couples could get married. The tunnel actually passed through a stunning crevasse, which was quite surreal. It was like something from a Disney film and was a major hit with the tourists. But for us there was little time for sightseeing. The weather had continued to deteriorate and visibility was also decreasing. The sky had turned an angry, deep gunmetal grey, with wind speeds increasing almost minute by minute and causing even Addi's massive truck to sway from side to side.

Even as we began to unload our equipment, we were being buffeted by the winds, but the bravado within the group was truly impressive – as far as we were concerned we were soldiers and big-time polar adventurers. I became slightly unnerved when one of Addi's assistants emerged from the ice cave. He had just been chatting with Addi when he walked over to me and said: 'I would like to shake your hand . . . because I think this is the last time I will ever see you.'

It was getting dark, so I decided the best course of action would be to remain in the area of the drop-off point, get the tents up quickly and ski off in the morning. We waved goodbye to Addi and told him to come and collect us in five days. My last memory was of him driving away, shaking his head in disbelief.

All the tents were up and our kit was inside within about twenty minutes, which was impressive given the difficulty caused by the wind. Minute by minute, the wind speed seemed to be building, so that by around 10 p.m. it felt as if you were sitting inside a jet engine. It became clear that sleep was going to be impossible. The Hilleberg Keron tents are designed for the worst conditions on the planet, but they were now being tested to their limit as the wind piled spindrift against them. The poles were under serious strain, so I decided that to provide a bit of protection we needed to build some snow walls, upwind, to shield the tents. Otherwise we were at risk of damaging the gear. Everyone was outside in the pitch black frantically digging, trying to construct walls in the howling gale, but we still had time for a few laughs and photo opportunities.

Strong winds blew all through the next day, keeping us pinned down inside our tents. We only ventured out to rebuild the snow walls and to check the guy lines were still firmly embedded in the ice. By late afternoon most of our equipment was soaked. With the added wind chill, our mini-expedition had turned into something of a survival mission, and the following morning I conceded that the weather had defeated us. After several hours of digging, we collapsed all three tents and took refuge in the nearby ice cave.

The simple act of escaping the howling wind was blissful, but it also dawned on me that we were achieving nothing sitting in the cave, and the best plan now was to get off the glacier, regroup, get our kit sorted out, grab a good night's sleep and hopefully return the following day.

I contacted Addi and asked him to collect us, but he was unable to come all the way up to the ice cave due to the weather, so we agreed to meet a bit further down. I briefed the guys that we would walk in single file, wearing crampons and pulling our pulks, to a rendezvous point at the bottom of the glacier. The hurricane-like winds had mercifully subsided enough to make the move down the glacier relatively comfortable. I led the way, but as we moved down quite a steep section of the glacier, Alun's pulk somehow came unclipped from his harness and began shooting down the glacier like a bobsleigh.

'Pulk,' Alun shouted. 'Pulk! Pulk!'

As I looked to see what all the shouting was about, Al's pulk shot past me like a rocket and headed straight for the edge of the glacier's cliff face to our flank. I immediately broke into an awkward sprint in my crampons and with my pulk still attached. Just as Al's pulk neared the edge, I dived and somehow caught the back edge with my fingertips, holding on to it for dear life until the others arrived a few seconds later. It was like a scene from *The Italian Job* as the pulk teetered on the lip of the cliff.

'Great save,' shouted Alex as he pulled the pulk clear of the edge. We sat around laughing for a few minutes, but the reality was pretty serious: had the pulk gone over, it would have fallen hundreds of feet down onto rocks and been virtually impossible to retrieve. The consequences for the expedition – and Al's part in it – could have been devastating. All of Al's equipment, his skis and one of the tents were contained within the pulk, and everything was due to be airfreighted to Chile a week later.

While some of the kit could have been acquired in the interim, it would have been impossible to get a new carbon pulk built in time.

After a night in a hotel, the weather had calmed, and we went back up to another nearby glacier where we spent the last day practising crevasse rescue drills. I showed the guys how to put in ice screws and build a ground anchor, rope up and set up a pulley system to give mechanical advantage. I explained the procedure for hauling people out of a crevasse and, when everyone was happy, one by one we lowered each person down into a huge crevasse, around 50 feet deep, before hauling them back out again. Each rescue was videoed as a training aid. I made sure everyone felt completely confident in what they were doing before calling it a day.

Later that night we had one last celebratory drink, which ended with a game of spoof, and this time the loser's forfeit was to buy a bottle of champagne – an expensive undertaking in Iceland. Alex lost and we followed him to the bar.

'How much is your cheapest bottle of champagne?' Alex sheepishly enquired.

'This one for £165,' the barman responded, offering Alex the chance to pay in sterling. Alex turned pale while the rest of us burst out laughing. Needless to say, Alex was gutted. I think it was that night someone first coined the phrase 'BEER17'.

The team returned home with only one thought on our minds: Antarctica was now in touching distance.

11

LEADER'S LEGS

The journey of a thousand miles begins with a single step.

LAO TZU

I watched the Twin Otter climb effortlessly into a perfect aqua-blue sky, scarred only by a few high-altitude clouds, while a gentle polar wind tumbled across the frozen Ronne Ice Shelf, where the ice is up to 1,000 feet thick and has remained unchanged for over 10,000 years. We were at Hercules Inlet, the start point for the expedition. The temperature had settled at a comfortable -20°C and the visibility was perfect.

I took in a deep lungful of cold air. For the first time in months I felt completely at ease, and caught myself grinning like a child at Christmas. It was so good to be back in Antarctica, despite my insistence at the end of the last trip that I would never come back. It is like no other place on earth: pristine, raw, unpolluted; even the light is extraordinary. A pure, perfect light everywhere. I felt as if I could see to the far end of the continent. I took another deep breath and smiled; the omens were good.

In that moment I thought of Henry and how proud he would be of what we were attempting to achieve. On our previous expedition together, he had written in the snow, 'I am Antarctica,' and in that brief moment of elation I knew what he meant.

We had spent the last two weeks facing delay after delay. First in southern Chile at Punta Arenas, and then at Union Glacier, as we patiently and then frustratedly waited for a weather window that would allow ALE to fly us across the continent to the expedition start point. Poor weather – high winds and whiteouts – had cost us two weeks. The expedition summer season in Antarctica is very short, and in our case was going to end on 24 January 2017, whether we had completed the expedition or not.

But nothing could stop us now. Finally, after all the planning, fundraising, training, and the seemingly endless waiting around, the expedition was about to begin.

Each team member checked their individual pulks to make sure that they had their kit where they wanted it and, most importantly, that the fuel hadn't leaked from the lightweight plastic canisters after the pulks had been double-stacked on their sides in the plane.

'OK, everyone ready? Cool, line up side by side,' I said to the team members, who were smiling broadly. I had complete and utter faith in their capabilities and determination. They were fit and motivated and I was immensely proud of them.

As Shackleton had done over a hundred years earlier, I shook everyone by the hand and wished them luck. It was an emotional moment. After the formalities of a handshake, we gave each other a hug. It was a real coming-together. Right then I had no doubt that we would complete a traverse of the continent, even though it was a monumental undertaking for a group of polar novices.

The guys were like greyhounds straining at the leash, desperate to get on the move, to begin breaking trail.

'To settle us in I will lead the first leg. I'll keep it nice and steady, and we will stop every hour for five minutes. Then we will change over, hand over the compass, and the second in line will lead and the previous leader will go to the end of the line. The changeover will be the only time we will stop as a team. If you fall behind, you will need to catch up at the changeovers. Everyone happy with that?'

And so, at 1600 hours Chilean time on 15 November 2016, we skied off from the start point, climbing gradually, almost imperceptibly, across the great Antarctic ice sheet, a frozen world where nothing lives. No trees or birds, no people or insects. Nothing for hundreds of miles except frozen ice and, somewhere in the distance not yet visible, towering, snow-covered mountains.

After around three miles we came across a tent belonging to Emma Kelty – or Tam, as she liked to be known. She had flown into Hercules Inlet on the same morning, but had insisted on flying on her own as part of the experience of her solo expedition to the Pole. We didn't stop to say hello and instead veered away to give her some distance. Tragically, Tam was murdered by pirates ten months later while attempting to kayak solo along the entire length of the Amazon.

We stopped at around 8 p.m. on that first evening and, although everyone was still buzzing with excitement, it was also dawning on us all how demanding the expedition was going to be. Each of us was carrying fifty days' worth of food and fuel. The pulks were at their heaviest, around 120 kilograms, but we were also at our strongest. Even so, after that first, short day, everyone was exhausted. It gave us all an indication of the task Henry had attempted, carrying food and fuel for eighty days.

The team quickly established the routine that we planned to follow every day for the next six or seven weeks until we reached the Pole. I was sharing my tent with Ollie, whom I'd already nominated as the expedition second in command. It was important that we shared the same tent so that he had a complete grasp of my planning and current thinking. If I became a casualty or incapacitated, he would be able to take over seamlessly. He and I also got on very well, which was vital.

Just as with Henry five years earlier, the evening routine began with pitching and securing the tents, followed by melting snow to rehydrate our meals and prepare our energy drinks. Journals were then written, kit repaired, and every evening I would nominate a member of the team to deliver an audio blog of the day's events, recording a voicemail message on the phone of Wendy Searle, the expedition media manager back in the UK. Wendy had very kindly volunteered for the role after hearing about the expedition during a meeting at The Soldiers' Charity, where she worked at the time. It was a huge commitment, involving several hours' work every day – in addition to her job and looking after her four children. The following morning, Wendy would place the audio message on the SPEAR17 website, as well as a full transcript of the recording across other social media platforms. That way friends, family and the media could follow our progress. Meanwhile, I would speak to ALE's operations room at Union Glacier and provide them with a situation report (sitrep). The three key pieces of information they required were the team's position in latitude and longitude, how far we had skied, and how long we had skied for. After that I would update them on any medical or kit issues – no matter how minor – so that they could prepare for any contingency actions.

My diary entry for that first evening read:

We skied 5.9 nm in four hours at an altitude of 1,675 ft. Dropped off by Twin Otter at 1500 hrs Chilean time. Short forty-minute flight from Union Glacier. Bit of turbulence and Ollie felt really sick and was sweating like mad. Weather was beautiful and no wind. Skied off and as we climbed wind picked up to a 20 mph headwind. Went past Tam's tent at base of climb. Found last hour difficult and felt really tired once inside tent. Used vapour barrier liners but they made heels sore. Chris got a blister on his toe.

I woke at 7 a.m. the following morning, thick-headed and sleepy. I gave Ollie a shake and in silence we got the cooker going and began the process of melting ice for breakfast. I had forgotten how grim mornings were in Antarctica. The willpower required to leave the warmth of a thick, down-filled sleeping bag should not be underestimated. Even on a good day, the temperature inside the tent was around -5°C before the cooker got going. The morning routine followed that of the evening, with the monotonous process of melting snow and preparing hot drinks and breakfast and hot water for our flasks for the day.

By 9 a.m. everyone was ready to set off for the first leg. I asked James to lead and told him to keep the pace steady. I wanted to avoid what the military call 'leader's legs', when the leader sets off at a really fast pace. This can happen at any time of the day, not just the start, even after the individual had been struggling to keep up with the pace when further down the line. It's as if the leader gets a new lease of energy. The guys were very fit and highly motivated, and had the potential to really push the pace. At forty-seven I was almost twice the age of Alex, James and Ollie, who were in their mid to late twenties, while Chris was thirty-four and Al was forty-two. This was definitely a marathon and not a sprint, and the key for me was to dictate a pace that we could sustain for two months.

It was also always in the back of my mind that I needed to have some spare capacity, just in case I had to make some important judgement calls. I never wanted to be at my physical limits and on the bones of my arse, or completely mentally exhausted, because if I was struggling to look after myself, then I wasn't going to be able to make the best decisions as a leader. This was something that I had learnt in the SAS. As a leader, especially on operations, it is vital that you never put yourself in a position where you are mentally incapable of making crucial decisions. This can often mean that others in the team may have to take a greater share of the physical burden so that the leader can remain mentally fit. It is essential that a leader always has a bit of spare capacity in reserve, so when things go wrong – as they almost inevitably do – they can step in and create a plan and manage the situation. So it was vital to know your own limitations.

Although the second day was sunny, we had to battle against a strong wind as we climbed higher and higher up towards the polar plateau. I noticed that by the end of the second hour Al was struggling with the pace and had dropped back a bit. At first it wasn't too noticeable, but as the hours passed he slipped further and further behind. The rest of the team didn't stop or slow down, and Al managed to catch up at the changeover points. By the afternoon, though, he had taken a turn for the worse and began to vomit.

I didn't want to make too much of a big deal about Al's situation, but I quietly asked how he was doing.

'Just finding it hard work,' Al said. 'I found yesterday a bit tough going and I thought today would be better.'

I had previously asked Ollie, who was looking particularly strong, whether he would be willing to take out a bit of weight from Al's pulk if needed, and he was happy to do so.

'OK,' I added. 'No worries, but this is what I want to happen. Ollie is going to take one of your food bags for the rest of the day just to take the strain off.'

Al nodded but remained silent. I made sure he took on some food and water, and five minutes later we were on the move again.

Al was very quiet by nature, and I knew that he wouldn't want to show any sign of weakness so I didn't pry. But by the middle of the afternoon, Al had slipped back again and was now around 300 yards behind the main group. When he caught up, I asked Chris and James both to take another of the 12-kilogram food bags.

'I'm OK,' Al said, almost in protest, but I was insistent.

'Don't worry about it, Al. It's no big deal – this is a team effort. It doesn't matter how we get to the Pole as long as we get there. The most important thing is that we are all travelling at the same speed and we are all comfortable. I know you want to carry all of your own weight, but you are clearly struggling with the pace and it's not safe to let you fall back too far.'

'OK,' was Al's only response, leaving me unconvinced that he had accepted my plan was necessary.

'At the end of each day, we should all feel equally exhausted,' I went on. 'If you finished the day way more exhausted than the rest of us, that is only going to snowball; you're going to be hanging out more and you'll deteriorate very quickly. Look, you might be fine tomorrow, so you can take all the weight back again – let's see how it goes.'

Al's pulk was now 35 kilograms lighter than it had been at the start of the day, but it made little difference and he still struggled.

I wasn't too concerned with Al's fitness; after all, anyone can have a bad day and I had plenty of them back in 2011 when I was with Henry. But I did wonder whether he had become daunted by the challenge ahead. You have to let Antarctica into your heart; you have to accept what it is and what it is going to demand from you without worrying about it too much. It can play tricks on the mind and make you fearful, and looking at Al I wondered whether this was happening with him. He had exactly the same symptoms that Lenny had exhibited on the Scott–Amundsen race: dry retching and loss of appetite together with a very pale complexion.

As we skied into late afternoon, I could see the Three Sails in the distance – a rocky outcrop and the first obvious landmark. Even though I had been monitoring our route using compass bearings and the GPS, it was still reassuring to know that we were on the correct course. I could see that Al's condition wasn't improving so I decided to make camp at 1800 hours. As per the previous day, it was straight into the evening routine of erecting the tent and melting snow.

That night, as Ollie and I settled down, he asked, 'Is Al OK?'

'Yep, he's fine. He might be slightly overwhelmed about the expedition. It does happen – I've seen it before. But it's early days. I don't think there is anything to worry about yet, but it is just something we all need to be aware of. Al's a robust guy. We all know he's quiet and he's never going to complain. There is plenty of time for him to bed into this. Probably a case of beginner's nerves.'

'Good,' Ollie said. 'I thought it might be something like that.'

The following morning we set off at 9.10 a.m. – ten minutes later than the agreed time. Ollie and I were ready to go, which meant that we were hanging around outside while the two other teams packed the last pieces of their personal equipment into their pulks. I made a mental note to remind them later that evening that a 9 a.m. start meant a 9 a.m. start. One of the biggest causes of arguments on expeditions is team members not being ready to move at agreed times. Because we skied in relatively lightweight clothing, being left waiting at -20°C was not OK. After just a few minutes of inactivity, Ollie and I were shivering and our hands had become stiff with the cold. In a virtually pointless attempt to keep warm, we began to impatiently windmill our arms to signal to the rest of the team that we were ready.

James led the first leg of the morning and immediately set a brisk pace – probably a bit too quick, as almost immediately the team began to separate. In theory the leader should glance back every few minutes to make sure the team was together and moderate the pace accordingly. I was number two in the line, so I increased my pace, caught up with James and just asked him to ease off until the rest of the team caught up.

After the second hour, Al had fallen behind again. The five of us who had kept together turned our backs to the wind at the short break, which meant we were all looking in Al's direction, munching away on our grazing bags while we waited for him to arrive. No one bitched or complained – if anything the rest of the team just expressed concern and, like me, were baffled as to why Al was finding the going so tough. It must have been pretty demoralizing for him skiing towards us, hanging out, finding the going really tough from the start of the morning, while the rest of us were operating within our comfort zones, looking fresh and enjoying the challenge.

Al finally arrived and dropped to his knees, looking exhausted.

'How you doing, mate?' I said, trying not to make too big a deal of what was now becoming a worrying situation.

'Not too bad, Lou,' Al responded, putting a brave face on things.

'It will get better,' I said, 'Just try and relax a bit and don't worry too much about being off the pace. I know what you are going through. I've been there myself, but I can assure you it will get easier.'

Al looked up at me but said nothing. He pulled out his grazing bag, threw a few handfuls of nuts and chocolate into his mouth and took a swig of water.

I sensed that the others were impatient to get moving again. They were probably now getting cold, but it was important that Al had his full five minutes' rest.

'It's just an early blip,' I told Al. 'One of those things that can happen to anyone. I fully expected a few issues with the team, so try not to let it get to you. We will all struggle at some stage and this is a team effort.'

The rest of the day continued in the same fashion, but we managed to cover over 12 nautical miles and skied for a total of nine hours. It was another beautifully clear Antarctic day. The sun shone brightly, and I passed the time listening to music and extracts from the masterful audiobook *Rogue Warriors* – a history of how and why the SAS was formed during the Second World War. I found the best way to get through the miles each day was to lock yourself into something, almost a distraction from walking across the endless frozen landscape. I would vary my routine each day, or sometimes by the hour, perhaps listening to music for a few hours, then switching to an audiobook, or sometimes I would just marvel at the majesty of my surroundings, listening to the ice crunching beneath my skis or a polar wind gently whistling in my ears.

Although we moved as a team, we only chatted briefly when we stopped, so most of the time you were in your own mind and staring at the back of the head of the person directly in front of you. It can be a very isolating experience, so it is crucial to keep positive – and that is a lot easier said than done if you are struggling either mentally or physically.

Three Sails provided an easy landmark, so there was no need to check the compass as often as we normally would, and we actually got a sense of progress as it loomed larger on the horizon.

I noticed that after around three hours I was also struggling, but after lunch and by mid-afternoon, my energy levels seemed to have restored themselves and the going got a bit easier. I was also having a bit of trouble with the 'skins' – a piece of material attached to the underside of the ski, which allows you to glide forwards but prevents reverse slippage. After battling away with it for about half an hour, I cut away the loose piece of material and normal service was resumed.

Every team member was getting used to their different pieces of equipment, their boots and the environment. Nearly all of us had some issues with kit, or had developed hot spots and blisters from the footwear. It was all relatively minor stuff, and to be expected, but it needed to be monitored and dealt with just the same.

We made camp around 6 p.m. and went into our usual evening routine. After a dinner of rehydrated chilli con carne, rice, and a litre of protein recovery drink, I headed over to Al's tent.

'How you feeling, mate?'

'Tired,' Al responded as he tucked into his evening meal.

'I thought it went better today. I know you struggled with the pace, but you carried all of your own equipment until lunchtime and then we only had to take one food bag. That to me is an improvement.'

Al looked at me and smiled but he remained silent. As usual he gave nothing away.

'Just one last thing, guys. Tomorrow I want everyone ready to move by 9 a.m. Not five past or ten past – but 9 a.m. We need to be ready to move. Ollie and I were hanging around for ten minutes this morning and got pretty cold. So, let's be on it tomorrow.'

'Yes, Lou,' Alex and Al responded in unison, and with that I headed over to the other tent and gave the same briefing to James and Chris, who apologized for their lateness earlier today.

I was happy that Alex and Al were sharing the same tent. Although they were very different characters, they seemed to get on well. Alex, at twenty-six, was the youngest member of the team and was incredibly enthusiastic and positive about almost everything. He was opinionated, and sometimes headstrong, but he was also up for everything and willing to try anything new. He would throw his all into every problem, including dull and mundane tasks, such as repairing damaged equipment.

While most of us regarded kit repairs as a chore, Alex approached it as an intellectual challenge, something to pass the time. He was a brilliant and proactive team member, the sort of individual anyone would want on an expedition.

By Day 6 we had crossed 81 degrees south, and that meant a 'party in the Ritz'. Somehow my tent had been named the Ritz, after the living quarters on board Shackleton's ship the *Endurance*. We had decided as a team that every time we crossed a degree of latitude, there would be a tot of rum or whatever booze we chose. Each of us was carrying a litre bottle of alcohol – enough, we hoped, for the entire expedition. It was a great evening and very relaxed. The audio blog had been filed early, diaries written and kit repaired, so it was a chance to relax and talk over what had happened in the past few days and what to expect in the coming weeks.

I filled everyone's cup with an equal measure of rum.

'Hey guys, congratulations. Passing 81 degrees south is a significant milestone and we are well on our way to the Thiel Mountains.'

We were lined up three on either side, and as well as a little morale boost, the evening was also an opportunity for me to speak to everyone together. More often than not, when we were on the move, the wind was blowing, making conversation difficult. Even when we were settled into our tents at night, it was hard to hold a conversation with the team because everyone was busy with their evening routine.

I asked everyone how they were getting on and how everyone was coping with blisters – a minor medical issue that could quickly become serious. The blisters I developed on the second day had already become open, raw wounds that needed to be cleaned and redressed regularly. Sharing a tent with a doctor became very handy! The Ritz party was also an opportunity for the rest of the team to interact with Al, give him some support and to genuinely assure him that we were still very much a team, and to address any other team issues.

I was all for everyone having a bit of fun, but not to the extent of damaging vital equipment. Earlier that day Chris and James had been messing around at one of the stops. James had pushed Chris over and – as he fell – a ski pole had been snapped, irreparably. Now that we were all together, I wanted to raise the issue of equipment care.

'Guys, we've only got four spares, now three. We need to be very careful,' I said to the team. 'We can't afford to damage equipment.'

Despite my pleas for the guys to be careful with equipment, by Day 13 two more ski poles had been broken. Chris, who simply didn't know his own strength, had broken one, and the other had been left on the ice where it was skied over by a member of the team. It was time for a team chat 'without coffee'.

'These are expedition-jeopardizing mistakes,' I told the team.

'We now have one spare pole and we still have a very long way to go until we reach the Pole. If we break another, then we risk having a major issue. Skiing with one pole is a nightmare, and that will slow our progress down dramatically – skiing with no poles is almost impossible, so if we lose another two poles that could be it. If we have to request an aerial resupply before we arrive at the South Pole, then we could be looking at a bill in the region of 15,000 USD.'

Everyone was suitably admonished. I was determined to make sure that the guys really grasped the importance of the issue, and it was probably one of the few times on the entire expedition that I was deadly serious. I struggled to sleep that night, worrying about the expedition being jeopardized because of a bit of foolish horseplay. The thought of having to return to the UK and explain to my commanding officer and sponsors that the expedition had failed because team members had been messing around was the stuff of nightmares.

Over the next couple of weeks, everyone began to settle into a rhythm, though Al was still a cause for some concern. There were days when he seemed to improve dramatically, only to struggle and slow right down the following day. There were always minor ailments that everyone picked up, including sore knees, a bit of chaffing on the backside, and the ever-present problem of blisters.

By Day 15 Ollie had started to develop sores on his face and painful ulcers inside his mouth. Initially it wasn't too bad, and James, the lead team medic, fully expected the condition to ease in the next few days. But the opposite happened, and the blistering on his face and lips grew worse. The three doctors and Chris, the paramedic, were confused. Cold injuries were obviously expected, but it was the suddenness and their severity that surprised us.

By Day 16 Ollie was having trouble eating, and – though he

never complained – appeared to be in almost constant pain. I was growing increasingly concerned.

'What do you think it is?' I asked Alex.

'I'm not sure. Initially I thought it was a cold injury, but he's the only one suffering. I just wonder if he has an allergy to the fur around his hood.'

'Could be,' I responded. 'Let's remove that and see if it helps.'

The following morning I woke early and glanced over at Ollie. He turned and looked at me.

'Jesus,' I said. 'That does not look clever.'

Ollie's entire mouth was a mass of blood and pus, and his lips had almost fused together. Whatever his problem was, it had got a lot worse.

Ollie tried to speak and I offered him a drink. Gradually he prised apart his lips while moaning in pain. In the time-honoured way of the British soldier, I then remorselessly took the piss out of him for the next ten minutes.

Ollie was laughing and crying at the same time. I told him to get a grip, then went outside and spoke to James.

'Ollie's face is a mess,' I said. 'I'll let ALE know tonight, but I'm now quite concerned about his mouth becoming seriously infected. If he can't eat, he will go downhill very quickly. I obviously haven't said this to him. I've just spent the last ten minutes laughing at him.'

'Roger,' James said. 'I'll give him some antibiotics and patch him up. He's a tough lad. He'll be desperate not to slow us down, but it's something you might want to factor into today's planning.'

Ollie soldiered on and his pace never slackened, even though he must have been in serious pain. But he wasn't the only one suffering. Al was working hard, probably harder than anyone, but he was struggling to get into any sort of comfortable skiing routine. Unless things changed, I began to wonder whether he would make it to the Pole. At this stage I also began to seriously doubt whether he would be fit enough to complete the second, more arduous leg from the South Pole.

By Day 17, I decided the time had come to give Al a bit of a motivational talk. He seemed to be unravelling quite quickly, and I was convinced his condition was mostly psychological. Antarctica has a knack of getting under your skin if you let it. During each break he would collapse to his knees by his pulk, and would have to be encouraged to eat and drink. I think we were expecting a little more resolve, self-composure and grit from a fellow soldier.

'Al, I know you're having a really tough time, but the team are doing all they can to support you. I know it's hard but it's also worth remembering it's an absolute privilege for any of us to be here in this incredible place. This is a once-in-a-lifetime experience, and there're a lot of other people out there who would give anything to be here right now. It's also really important that once this is all done you can walk away and be happy with your conduct and know that you gave it your all.'

I explained to Al that if he gave up, he would regret it for the rest of his life. I knew from personal experience in the SAS that guys who accepted failure had to live with the knowledge that they allowed themselves to be beaten by a bit of pain and suffering.

'This suffering is only temporary and will come to an end,' I told Al. 'What's important is how you feel about yourself afterwards. I'm not sure if it will help, but on the trip I did with Henry I had inscribed on the front of my diary "Survive with Dignity".'

Al nodded in agreement and I think he got where I was coming from.

'I know what you're saying, Lou,' Al said, 'but this is another level. I just didn't think it would be this hard. It makes getting into the Reserves look like a picnic, to be honest.'

We both laughed, which was good.

'I know what you mean, but you will reflect back on this expedition for the rest of your life – I can guarantee that – and you don't want to be thinking: I didn't really give it my all. You want to be proud of your performance, knowing that you gave it your best shot. If not, then how you carry yourself after this will prey on your mind for the rest of your life. You will have to live with that. It's fair enough struggling, we all will at some stage, but you need to come through this with a bit of dignity and honour as well.'

Al seemed to respond well, and our chat really seemed to strike a chord – it was the sort of reaction I had hoped for.

He seemed to now be more energized and said: 'I'm sorry, Lou, you're right. I'm not going to give up. I'm really going to try my best – the last few days have been a bit up and down. I'm working 110 per cent.'

'That's great, Al, exactly what I wanted to hear. Tomorrow is another day so let's go for it.'

As I returned to my tent, I also sensed that by now a few of the other guys were getting a little bit frustrated with hauling some of Al's weight, although they never said anything to me. But under the surface I could see the start of a sort of slow, simmering resentment that I knew I would have to keep a lid on and manage. Antarctica had a knack for bringing out the worst in people. Scott, Amundsen and Shackleton had all experienced man-management issues during the course of their expeditions. Scott invalided Shackleton back to England on the Discovery expedition, triggering a simmering rivalry between them, and Amundsen removed Hjalmar Johansen from his final Pole party after the latter accused his leader of abandoning him during a false start towards their goal. Amundsen would go on to describe Johansen as 'violently insubordinate'. I was determined to stop any niggles escalating. There was a certain inevitability about the situation, and I think Al understood that as well. If someone was having a hard hour and they were hauling an additional 15 kilograms of someone else's gear, they would convince themselves that their problems were due to the fact that they were carrying that extra weight. In between those brief five-minute stops, most of the time on the ice is spent alone with your thoughts, and small issues, such as a beef with another team member, can gnaw away at you so much that by the end of the day you are hating that person.

The evening was capped off with another party in the Ritz because we had reached 84 degrees south. Despite Al's struggles, we were making excellent progress and, if all went well, we had a real chance of reaching the South Pole either on or just before Christmas Day. My chat with Al must have had some effect, because the following day he was fine and managed to haul all of his own weight for the entire day.

While most of the team were getting to grips with travelling on skis, Al and Chris were still struggling with the fine art of being efficient, which may have been part of the reason Al had found the going so tough. A good technique is key to conserving as much energy as possible, particularly on a long-range journey such as this. The idea is to just unweight the ski enough for the skin to break traction, and slide the ski forward without lifting it too much. It's more of a pushing and gliding motion than walking, although the glide is very limited when hauling a laden pulk. If you adopt more of a walking style you end up lifting the boot and ski, which combined are pretty heavy, and you will tire much more quickly. You are also stressing the binding and steel toe bar in the boot in ways for which they are not designed. Eventually something is going to give, as we were later to find out.

By Day 19 the surface was now effectively hard enough to walk on, and out of frustration Al and Chris decided to carry their skis on their pulks and try walking across the ice in their boots. I had hoped that almost three weeks into the expedition everyone would have developed an efficient skiing technique. While walking might have appeared easier, both Al and Chris finished the day significantly more tired than the rest of the team. Later that evening I told them both that walking was not a sustainable choice – physically they would both deteriorate at a quicker rate than everyone else – and in deep, soft snow, walking was impossible. I wasn't overly concerned about Chris because he was so strong and could probably pull two pulks if needed, but Al's decision not to ski caused me further worries.

The days now took on a familiar routine. Ollie's face continued to be a problem. James would dress his sores as needed, while the rest of us would take the piss out of him every time he moaned in pain.

Al had now accepted that he was going to struggle with the pace, although some days were better than others. The very fact that he hadn't quit despite suffering immeasurably was testament to the deep reserves of strength he clearly had. While the rest of the team woke every morning buoyed by their desire to get on the ice and ski for as hard and as long as possible, Al knew that all he had to look forward to every day was ten hours of mental and physical anguish.

I could tell that Ollie, James and Alex were beginning to get frustrated with the pace, and so at around 1700 hours on Day 21, I told them to push ahead while I waited for Al to catch up. The three doctors were the youngest and probably the fittest members of the team. They were like machines, and seemed to have an unnatural ability to eat up miles without any physical effect.

'Right, guys,' I said, pointing at them, 'I know you've been desperate to be let off the leash, so I want you to go as far as you can for the next two hours then stop and start getting the tents up.'

'OK. Will do. No problem,' Ollie responded with his usual enthusiasm. 'We will push on until 1900 hours then set camp.'

I smiled inwardly as I watched them excitedly ski off into the distance, pushing one another to see who could maintain the fastest pace. As the three of them disappeared into the empty whiteness, Al, Chris and I set a more moderate pace, skiing in the tracks of the three racehorses.

By the time we had reached the campsite, we had covered more than 17 nautical miles – one of the highest mileage days of the entire expedition. The tents were already up and the evening routine of melting snow was already well advanced.

As I checked in on all the tents later, I realized that James, Alex and Ollie had really enjoyed what had effectively amounted to a two-hour polar sprint in which they had tried to out-ski each other. The following morning, however, they all commented on how tired they felt. Their legs were stiff and they felt that they had overexerted themselves. I was quietly pleased, as I think they realized for the first time that I was setting the correct pace, even if they occasionally felt it was too slow. I reminded them again that we needed to start and finish each day as a team, and that it was important we all felt as though we had equally exerted ourselves. It was never going to be an exact science, but they got the gist of what I was trying to achieve. From then on there were no calls by anyone to go faster.

That night Alex and Chris swapped tents. Al felt that he was losing sleep because of Alex's snoring, so Chris moved in with Al while James and Alex teamed up together. It was all sorted very amicably. Ollie and I were getting on great and keen to remain tent-mates throughout.

The following morning the Thiel Mountains' jagged peaks, which had remained above the snow line for aeons, loomed large on our right-hand side. We were now skiing at 4,000 feet.

The team gathered inside my tent after dinner. One by one each team member brushed snow off their boots in the vestibule at the end of the tent, before heading into the inner sleeping compartment. It was cramped but cosy.

'Just wanted to bring everyone up to speed with where we are,' I said as I unfolded part of a map.

'We are now pretty much halfway to the Pole.'

Everyone gave a little cheer before I added: 'Which means we have more than 350 miles to push. Whether you are aware of it or not, we have all lost quite a bit of weight, both fat and muscle. Health-wise, although we all look in good shape, we are beginning to run down. So while our pulks are getting lighter gradually, we will be becoming physically weaker.

'So, what does this mean? Well, it means teamwork is now even more crucial. It's been spot on so far – everyone helping out, especially when Al has been struggling – and I can assure you that we will all need help at some point before this trip is over. We are also now climbing each day, and we will continue to do so until we reach the top of the polar plateau at around 9,300 feet. It's going to get challenging, but I've no doubt that we will make it.'

The morale in the team was sky-high, and everyone was looking forward to the next phase of the expedition. The following morning we were up early, and set off just before 0900 hours into a strengthening headwind, which lasted for several days. The mountains remained in view to our flank for the next few days, gradually slipping further and further behind and then disappearing as we headed south. It was refreshing and comforting to know that in such a busy world, no one had come close to setting foot on any of the summits.

We would all listen to music or audiobooks as we skied along, which were great for drowning out the persistent wind noise and keeping the mind occupied. However, Alex's music collection had come courtesy of Apple Music and, unbeknown to him, after thirty days of his phone having not connected to the internet to reconfirm his subscription, all his music froze. All he was left with was an audio recording that a close friend had made for him, which he decided to keep for a low moment. When the day in question arrived and he felt he really needed a boost, he eagerly hit play on his phone, hoping for some inspirational words to help lift his spirits. Instead he was treated to his friend and his girlfriend reciting a particularly racy chapter from *Fifty Shades of Grey*. At least it provided us with a great laugh when we found out later.

On Day 28, Alex developed an unusual condition called polar penis. It might sound funny, but it's an extremely painful cold injury affecting the groin area. Alex was leading the team into a strong headwind and, as the guy in front, he was taking the brunt of the gusts. It was like standing in front of a giant fan blasting out cold air at a temperature of around -40°C. To put it simply, he developed a super-chilled and extremely painful todger. The pain was partly alleviated by stuffing a woolly hat down the front of his trousers.

Later that evening, after we had set up camp and had dinner, it was Alex's turn to deliver the audio blog, and he explained to the outside world how he had been suffering from polar penis for most of the day. The newspapers got hold of the story and, under the headline, 'FREEZE WILLY', the *Sun* told its readers that 'Antarctic explorer reveals he's suffering chilling condition known as "Polar Penis" with nether regions in agony'.

The *Sun* made quite a big deal of the story, and within hours rumours began to circulate that Alex had actually lost his penis due to frostbite. Alex's mum, Kate, was blissfully unaware of her son's predicament, until she was stopped by a friend in the street who said: 'Kate, I'm so sorry.'

Kate, confused by the comments, replied: 'What? What do you mean? What's wrong?'

'I'm so sorry to hear that your son has lost his penis in Antarctica.'

Mortified, Kate dashed home and asked her husband what had happened, only to be reassured that nothing had been detached or fallen off, and Alex was still all man.

While Alex was struggling with polar penis, James was struggling to maintain his body temperature. Unfortunately, despite using all his mid-layer clothing, he ended the day seriously cold.

He was lethargic and nauseous, but it is always difficult to spot the early signs of hypothermia – there was no shivering and the main difference was that he was decidedly grumpier than his usual upbeat self. It was a salient lesson for all of us as we were reminded how, despite being relatively acclimatized to Antarctica's notoriously turbulent weather, danger was never far away. Body temperature regulation was probably one of the most crucial skills for survival here. There can be a high price to pay for getting too cold, but likewise sweat too much and the moisture will freeze quickly and cause serious problems as well – between those two extremes was a thin range, within which we consciously attempted to remain. 'Comfortably cool' was the mantra I had inherited from the highly experienced polar guide Hannah McKeand.

The following day James's condition deteriorated and he was struck by a bout of diarrhoea. It slowly became apparent to the doctors that James was suffering from a severe infection that he had developed after getting a blister on the side of his foot. His boots had been rubbing away at the skin for some time and, although he had dressed the blister every night, the wound had become infected. He had attempted to cut away the inside of his boot, in the same way that I had done when I was in Antarctica with Henry back in 2011, but it appeared to make little difference. Also, the now rundown condition of our bodies meant that fighting off infections was harder.

To make matters worse, we entered a period of whiteout, which was mentally challenging at the best of times. By 1300 hours the next day, Alex, Ollie and Chris had taken about 30 kilograms of weight from James's pulk. We were now moving so slowly that I was worried about James or other members of the team getting hypothermia. Up until that point, James had been one of the strongest in the team and I was stunned by how quickly he began to struggle. My words of warning just a few days earlier now seemed to be alarmingly prophetic.

James 'ninja'd' it out until 1700 hours when he turned to me and said, 'I've got to stop.' I immediately knew we had a problem.

Despite all of the tough days we had endured, no one had actually ever asked to stop. In that fleeting moment, as James appeared close to collapse, I realized we had all missed the signs that his condition had been deteriorating for a few days and now he had gone into freefall and could barely stand.

I hadn't spoken to James for an hour because he had been ahead of me in the line of march, but when I saw his face I knew he was now in a serious condition. He was pasty-white and his eyes were distant. He was just about holding it together, and my first thought was shit, he's going down. I had seen the look before on selection, both as a candidate and an instructor. Some soldiers seem to have a capability to push themselves to extremes, but when they reach their limit the body quickly closes down, as if a switch has been flicked in a futile attempt to preserve the last of its strength.

We immediately leapt into action and began to get tents up and cookers on to melt ice so that we could make improvised hot-water bottles with our Nalgene containers. We got James into a sleeping bag as quickly as possible and then packed the Nalgenes around his body. His core temperature had become dangerously low. He had made a good call asking to stop, and it showed he still had his wits about him to turn to me when he felt he needed to.

James was in a bad way; he was lethargic and – to make matters worse – the whiteout meant that any form of casualty evacuation was impossible. But restoring his body temperature had an amazing effect; there was a rapid improvement in his symptoms, and I left him in his tent with Alex feeling slightly better. The best we could hope for was that he would turn the corner and start to fight off his infection. Gradually, over the course of the next few hours and despite being very concerned for James's condition, I grew more confident that James was going to pull through.

Later that night I got James to call his parents, not because I was worried about his survival, but because details of his incident would have appeared on the blog, and I didn't want another panic-inducing 'polar penis' situation.

Fortunately, with Ollie and Alex and Chris – two doctors and a paramedic – still fit and well, James couldn't have been in better hands.

The expedition was temporarily halted for the next twenty-four hours while James recovered. He was placed on antibiotics, slept for most of the time and ate.

James's near-miss was an important lesson for all of us. I reinforced the message that we had a responsibility to ourselves and to each other to recognize the early signs of fatigue and hypothermia.

We made the most of the unexpected twenty-four hours' rest and played cards, conducted much-needed kit repairs and slept and ate. It was one of those occasions when, given the opportunity to stop, you discover how tired you actually are. Remarkably, just twenty-four hours later, James was back on his feet and we were on the move once again. Within two days he was looking strong once more.

Any ideas we had that the rest would reinvigorate us were misplaced. The following day was really hard work, and everyone struggled to keep up the usual pace. It was almost as if we had developed a rest hangover. But over the next couple of days we settled back into the usual routine. Al was still having problems keeping up, though we were moving so slowly that the rest of the team were getting cold, so I thought I would have a go at double-hauling. The two pulks were attached one behind the other and I fixed my harness to Al's pulk so we could haul side by side and chat along the way. The change was immediate, and the team was soon travelling at a decent pace, which allowed everyone to keep warm. It was further proof – if any was needed – of the mental demands of an Antarctic expedition.

'Bit easier, isn't it, skiing side by side,' I said to Al, who was working at his full capacity.

He looked over at me and smiled.

'Certainly is. I know everyone's probably a bit fed up with me. I just didn't think I would struggle like this. I'm doing my best,' Al said, clearly concerned about how he was being viewed by the rest of the team.

'Don't mention it, mate. Everyone knows that. So honestly, don't worry about it. What we want to try and do is pass 88 degrees south today – then it will be a party in the Ritz tonight and then we'll be on the home stretch. Just two degrees left until we reach the Pole. What do you think? Can you push yourself that far?'

'Of course. I'll give it my all,' Al replied.

On we pushed, ticking off the miles, climbing higher and higher, everyone blocking out the pain and the fatigue until I got the GPS out, which showed that we had crossed 88 degrees south. It was Day 34, 18 December, and we had covered around 580 miles with another 150 miles to go until the Pole. Every degree is equivalent to 60 nautical miles and later that night, as we partied in the Ritz, I raised an idea with the team.

'OK, everyone. Congratulations on getting to 88 degrees south.'

Everyone cheered and raised a glass of malt whisky.

'So, I've got a proposition for you all. What do you say to the idea of skiing into the Pole on Christmas Day? Before you all answer, to achieve that we will need to cover 120 nautical miles in seven days. That is just over 17 miles a day – more than we have been doing. We are on the polar plateau now, so the going will be a bit easier. I want everyone to be completely honest with themselves, because we only do this if it has unanimous support.'

I let the thought settle for a few seconds, then the heads began nodding and Alex said: 'Arriving at the South Pole on Christmas Day – that would be super-cool. Imagine that.'

'Yeah. Imagine that: walking into the Pole on Christmas Day – once-in-a-lifetime achievement. Gotta be done,' added Chris.

One by one, every team member chimed in. Al was quiet, as ever, and I didn't want to single him out. He nodded his approval, but I knew that privately he was probably cursing me.

We all had a spring in our step as we collapsed the camp and pushed off by 9 a.m. the following morning.

My thoughts soon turned to the next and possibly the hardest part of the expedition from the South Pole to the Ross Ice Shelf. It would mean skiing up and over the Titan Dome and down through the Shackleton Glacier, a part of the journey that Henry had found really difficult. The big question I had to ask myself was, could Al make it? Did I take him along or give him the opportunity to withdraw?

I decided to seek Ollie's advice. Ollie was one of the toughest people I had ever met. He seemed to have a unique capacity to absorb pain, as his ability to cope with a now clearly frostbitten face had shown. He had a fantastic belief in his physical abilities, and I honestly believed that he felt there was nothing physical he couldn't achieve without the right preparation. It wasn't arrogance, he never boasted about it – it was a form of mental strength.

'Do you think Al should be on the next phase of the expedition? We'll have time to rest and patch ourselves up, but we are going to be hauling full loads again. My aim was that we should all finish together, but I'm having serious doubts about Al,' I told Ollie.

He thought for a while then said: 'Look, Lou, I've been carrying a lot of Al's weight almost from the start. It hasn't been easy for me, and there have been days when I felt he didn't really appreciate it, but I know he's doing his best. He's a tough guy. He doesn't complain. He just gets on with it, even though he's probably finding it far harder than the rest of us.

'But I have to say this. I'm not willing to carry any of his stuff beyond the Pole. I think everyone has to accept that if you do the next phase, you have to haul your own gear, and that goes for everyone – you, me, Alex, Chris, James and Al. If you don't believe you can do that, then I say there is no place for you. But it's your call. You're the boss.'

Ollie was right. I was the expedition leader and it was my call. Had this been an operation, I would have removed Al without hesitation. Someone clearly struggling could have jeopardized the mission and put the lives of the rest of the team in danger. But this was an expedition, albeit a military one, and there was a difference. I had chosen and trained all of them. I had invested time, energy and trust in each one, and I wanted us all to complete the expedition. Then again – no one said being a leader was easy.

Over the next few days, other members of the team also voiced their concerns to me about Al and whether he was strong enough to complete the next phase. I was determined that Al shouldn't feel as though he was being pushed out of the group and so, although I listened to the others, I gave little away. Every team member was tired and we were all hauling relatively light pulks. Most of our rations and fuel had been used up but our bodies had taken a beating. We had all lost weight and were probably a little malnourished. At the Pole we would be collecting another thirty days' worth of food and fuel.

On Day 38 we crossed into 89 degrees south at around 1200 hours but, rather than stop, we continued bashing out the miles just to get some credit in the bank, if you like.

My diary entry read:

Absolutely amazing! Crossed 89 degrees south. Party in the
Ritz to celebrate. Team buzzing – three days to Pole now.

A three-day degree is a tall order. That's 60 nautical miles in three days – 20 nautical miles a day for a team that had already skied over 650 miles – but morale was high and we were all determined that nothing would stop us.

The following day was a major challenge almost from the beginning, and I briefly wondered whether we had burnt ourselves out. It was a long, cold day and our pace had dropped considerably. Every step was a major effort, and even Chris, who seemed to be afforded superhuman strength, was struggling to stay with the pace. At each hourly stop I checked on everyone, looking at their face, attempting to gauge how much they had left in the tank. I reminded them that there was an opportunity to get Christmas dinner at the Pole – it was a huge motivator but not at any cost.

The following morning, Christmas Eve, we set off early again at 8.30 a.m. The determination to get to the Pole the following day was now huge, and I don't think anything or anyone could have stopped us. We had collectively grown used to fatigue and discomfort. I personally thought of little else but of having a few beers, possibly some wine, fresh food and a wash – that was all I needed to motivate myself through the next few hours. By the time we stopped for the evening, we were just 12 miles from the Pole. That night we had a little Christmas party in the Ritz. We had decided back in the UK that we would have a team party wherever we were, and we had all bought Christmas presents for each other. Everyone had a party hat, there were lightweight crackers, miniature Christmas puddings, and I had a box of Ferrero Rocher chocolates, which was all shared out, as was the last of the booze. It was a great evening and I sensed that it might be the last we spent together as a full team.

Everyone was up bright and early on Christmas morning, and we spotted the faint outline of the South Pole Station at around eight nautical miles' distance.

'There it is, guys,' I said, pointing into the distance, 'that's the

South Pole. It's five or six hours away, maybe less. Well done everyone – you are members of a very small club.'

Had anyone been listening, they would have heard whoops of joy and seen six men dressed in blue jumping up and down, cheering and falling to their knees in tears.

It was a hugely emotional moment, as well as a major accomplishment, and I felt extremely proud of what the guys had achieved, but we still had some distance to go. As we skied off, we dropped into a large, dipping undulation in the ice, and the station disappeared; by the time we had re-emerged a whiteout had set in, and the Pole had effectively vanished.

At around 2 p.m. we emerged from the whiteout to discover that we were just a few hundred yards from the South Pole Station. We lined up abreast like the characters in *Reservoir Dogs* and skied silently in unison.

For the most remote place on earth, the Amundsen–Scott South Pole Station is actually quite busy. The main station is composed of a long, flat, graphite-grey building with a row of windows running along the front face. It is more reminiscent of a modern prison block than a scientific research centre, and has been built to withstand the most extreme climate on earth. Within the complex is a series of offices, laboratories, sleeping quarters and a dining room, a full-sized gymnasium, and a greenhouse where vegetables are grown in the summer months. Behind the complex is a vehicle park where dozens of ice tractors, Ski-Doos and specially adapted polar vehicles are parked. The entire operation is resupplied by air from aircraft that land on a half-mile-long runway composed entirely of ice.

The small ALE tented camp is located a few hundred yards from the South Pole complex and, as we approached, Hannah and Ricky from ALE and some other expeditioners came flooding out to congratulate us on our achievement. Tourists from China, Korea, and the United States were hugging us, slapping our backs and offering us high-fives. It was completely unexpected and a wonderful reception. In that moment we felt like explorers who had conquered something never before achieved. After the very quick obligatory photographs at the actual Geographic Pole, we headed into the long and crowded dining tent where Christmas dinner was being served. I was drooling like a hungry dog as I began to help myself to turkey and all the trimmings, stacking my plate with more food than I could possibly eat.

I turned to see the others all doing the same, their weather-beaten, suntanned faces staring at the food on display. They were all grinning like crazy men, laughing, joking and hugging complete strangers. Chris's plate was so full it seemed to defy the laws of physics, and I had no doubt that he would demolish the lot in a matter of seconds. The sheer joy of being able to eat fresh food after six weeks of a diet based on freeze-dried, rehydrated expedition food is something difficult to describe. I don't think any of us knew just how hungry we had become. As we tucked in, a member of the ALE staff began passing us glasses of wine and cans of beer. Our dreams were now complete, and the pain and discomfort that had been part of our daily routine for so long soon faded into the alcohol-induced fug.

As the evening wore on, our ability to stay awake began to decrease. Alex fell asleep in front of a plate of food, probably his third dinner of the afternoon. We pitched our tents and crashed, sleeping in late, delighting in the knowledge that we could sleep off the hangovers and rest without having to haul ourselves across miles of snow and ice.

On Boxing Day, the team were given a tour around the South Pole Station by the US scientists, and later that day each of us was given a medical check by James and the ALE doctor. It was apparent that we had all lost a great deal of body fat, and some muscle mass, but Al had lost the most, which partly explained why he had struggled throughout the expedition.

I was planning to interview each member on the 27th before we set off again for the second phase of the expedition on

28 December. As far as Al's future on the expedition was concerned, it was decision time. I had held off until I had the results of the medicals and the weigh-in, as well as some advice from the very experienced ALE staff who had seen hundreds of expeditioners arrive into the South Pole in varying states. They had noticed a marked difference in Al's appearance and demeanour compared to the remainder of the team. He looked like a broken man.

The team interviews began on the afternoon of 27 December and Al was first on the list. I had his medical report in my hands, but I began first by asking Al how he felt the expedition had gone so far.

'I've really enjoyed it,' he responded with a smile. 'But physically, on most days, I was operating on another level to anything I had ever done before. I know you said it was going to be hard, but I don't think any of us expected quite how tough the journey was going to be. I was always working flat out just to keep up. I really appreciate that the rest of the team helped by carrying some of my weight – that made a huge difference.'

'It's no wonder you found it so hard,' I said, looking at the medical notes. 'You've lost 13 kilograms in weight. That's quite a lot for the first leg. I think it's a reflection of your determination that you reached the Pole. But I now have to consider what happens next.'

Before I could continue, Al interrupted.

'I've already thought long and hard about this, and I don't want to slow down the team. I really wanted to complete the second half of the expedition, but I'm not sure it's meant to be. The next leg is going to be harder, particularly with the extra weight of the resupply, and to be honest I'm physically exhausted.'

I was deeply impressed by Al's honesty and moved by his words. It was a perfect example of the selfless commitment and humility that underpins the values and standards of the British Army. Just as Captain Oates had sacrificed himself over a hundred years earlier in the belief that his frostbite was putting the rest of the party at risk, Al was sacrificing his place on the expedition for the sake of the team.

'I don't know what to say, Al. That's an incredible gesture.'

'There's nothing to say, Lou. The expedition has been one of the hardest things I've ever done, but I wouldn't have missed it for the world. I have no regrets.'

'You've gone through hell, Al, and done well to get this far, and I really appreciate everything you've done, particularly in the build-up to the expedition. I'll make a statement that for reasons of safety, due to your excessive weight loss, I'll be removing you from the expedition at this point.'

He looked visibly relieved at the decision and stood up, shook my hand and left the tent. Over the next hour or so I interviewed the rest of the team, explaining that Al would be leaving the expedition at this point. Everyone was seriously impressed with his attitude and I knew that we would all miss him.

Ollie's chat with the doctor confirmed he had suffered a cold injury that amounted to a form of frostbite. It was good news that his condition had finally been correctly diagnosed, but the bad news was that it was unlikely to improve until we had finished the expedition and moved him into a warmer climate.

The only other issue that gave me moderate concern was Chris's legs. His unusual skiing technique had a far greater impact on his legs than was normal, and by the time we had reached the Pole, his quad muscles were seriously fatigued. He had really struggled on the final couple of days and I just wondered how much he had managed to recover. Time would tell.

12

THE TITAN DOME

There's no such thing as bad weather,
only unsuitable clothing.

ALFRED WAINWRIGHT

I could see that it was daunting for the team to leave the creature comforts of the Pole behind and once more head into the polar wilderness. But personally I was pleased to be on the move, heading away from all the distractions the Pole had to offer.

Our route would take us up and over the notoriously wind-swept, desolate and frozen wastes of the Titan Dome, which summits at an altitude of around 10,000 feet, then down through the Zaneveld and Shackleton glaciers and on to the Ross Ice Shelf. Although shorter than the first leg, the terrain was far more complex, and very few expeditions had ventured over this side of Antarctica.

Despite the challenges that lay ahead, the omens were good. The sun shone like a summer's day in England, the winds were light and morale was high. I was still concerned about Ollie's frostbitten face, which was now covered in various protective dressings. He looked like the invisible man, and it goes without saying that, as we checked out of the South Pole and said our goodbyes to the ALE staff, he was remorselessly ribbed. It was one of the lighter moments of a day full of mixed emotions. While it was good to be on the move again, we all felt a tinge of sadness that Al wouldn't be joining us. The team had grown close over the previous weeks, and it didn't feel right to be leaving him behind, but it was absolutely the right decision. He skied with us along the edge of the station airstrip for half an hour or so before we said our final farewells.

Then we were on the move again, in single file on a slow but steady incline up towards the Titan Dome, a huge, mysterious bulge in the ice. You feel more exposed to the winds out here and, even in summer, temperatures can plummet below -40°C. Henry's vivid blogs of the seemingly endless slog to cross it were fresh in my mind.

What lies beneath, no one seems to know. But for us the Dome meant two weeks of unrelenting hard work – possibly the two hardest weeks of the entire expedition – and the pain began immediately. The tethered pulks, restocked with 70 kilograms of fuel, food and equipment, tugged on our now bruised waists and shoulders, and very quickly the relative comforts of the South Pole faded from our collective memories. Each step, each slide of the ski through the sand-like, snow-covered surface required every ounce of effort we could muster. Everyone seemed to be struggling, even Chris. The South Pole complex shrank in size every time I glanced back over my shoulder to check on him, only to see him slipping further and further behind. Although he was known as the machine, the fact that he was now in such a bad way was a real concern, and I seriously questioned whether he would be able to continue. It was something we would have to have a frank discussion about.

By early evening the South Pole station had disappeared from view, as if it had been consumed by the vastness of the continent. Once again, we were on our own – the five of us and the Great White Queen. The sense of isolation in Antarctica is unnerving. There were times when I felt as if we were the only people left on the planet – specks of nothingness on an endless white landscape, unwanted visitors to the last unpopulated place on earth.

As the hours passed on that first day of the second leg, my

mind began to wander to my future. I was due to leave the Army on 20 February 2017, just three weeks after returning from the South Pole, and I had no idea what I was going to do. It was quite daunting.

I had spent the last few months putting so much effort into the expedition that I had forgotten about the need to find employment, and now I was hit by a mild form of panic. I had been in the SAS for twenty-five years. It was all I knew. When I first joined back in 1992, I felt that my career would go on for ever, like a young footballer who breaks into the first team – you never contemplate the end. Then one day you wake up and the end of your career is just beyond the horizon. Leaving the Army, and especially the closed, tight-knit world of the Special Forces, can be a difficult time. In a matter of weeks your life changes irrevocably. For a lot of guys, everything they know and love is gone – their home, if they lived in an Army quarter, job, support network and many of their friends. I also wondered what impact leaving might have on me and what I was going to do for work.

But there was little I could do about my life after the Army while still in Antarctica, and the last thing the team needed was a leader worried about his next job. I had to focus on the here and now, and the first issue to resolve was Chris and his worrying lack of pace.

I woke early the following morning and immediately knew that it was going to be a long, hard day. Ollie was still fast asleep, so I closed my eyes for another five minutes before crawling out of my warm, cosy down bag and firing up the cooker. The mornings never got any easier, and I soon began to shiver as I piled snow into the kettle. When Ollie stirred, I told him to remain in his bag for a while and passed him some hot chocolate. The low temperatures, howling winds and the prospect of another tough day on the ice quickly quelled any desire either of us might have had to engage in small talk.

By 9 a.m. we were on the move and Chris, who had barely uttered a word for the past hour or so as we broke camp, quickly slipped behind again. Alex had set a blistering pace, which I was also struggling with, and I was forced to ask him to slow down. The surface of the ice soon deteriorated. A layer of spindrift snow had hidden the deep, ankle-busting ridges of sastrugi, which was causing everyone to slip, fall and curse our surroundings. Sastrugi, the Russian word for snowdrift, are created by the katabatic winds flowing down from the plateau towards the coast, carving out incredible shapes in the snow. They are rock hard and can range from a few inches high to several feet.

As the day wore on, the Dome's incline seemed to steepen, but that was probably illusionary. I looked back after forty-five minutes and noticed that Chris had removed his skis, preferring to walk across the sastrugi. It was a further indication that all was not well with him. After the first hour, we stopped and waited for Chris to catch up. As he came into view, I could see that he was sweating heavily and a huge frosty icicle hung from his bearded face like a kitsch Christmas decoration.

'How are you doing?' I asked, while shielding my face from the icy wind.

'Struggling, but the last ten minutes were easier,' Chris responded.

'Why aren't you using your skis, mate? You need to be skiing. The terrain is going to get tougher. You need to concentrate on your technique and make it easier for yourself. You won't be able to walk for the next 400 miles.'

'I know, I know,' Chris added, clearly frustrated. 'I'll put them back on for the next leg.'

'Cool. I'll ski behind you and see if I can give you some pointers on your technique.'

'I'm getting a bit worried about slowing down the rest of the team,' Chris said, and I could see the concern ingrained in his bearded face. 'If I can't keep up then I'll take myself off. My legs feel like they have got no power in them. I can't ski and the walking is killing me.'

While Chris was having problems walking, Ollie's frostbitten mouth was weeping with pus and blood. His whole face was swollen and he could barely speak. When I asked how he was doing I got a thumbs-up, but I knew that he must have been feeling terrible.

After a five-minute break we set off again. I skied behind Chris, encouraging and cajoling him in equal measure, but most of what I shouted was lost in the icy wind. Just after midday Chris somehow managed to snap the steel toe bar on his boot, which connects it to the ski. He was already known as Conan the Destroyer because of the amount of kit he had broken, including skis, poles and bindings, but breaking the toe bar was something else. This was a serious problem. To put this into context, toe bars just do not break. They are one-inch thick rods of high tensile steel moulded into the toe of a specialized ski boot. In all my years of polar travelling, I had never heard of a toe bar failing. But Chris managed to snap his because of his poor skiing technique. He was lifting and stressing the gear with every step and had been doing so since day one. Most galling was that Chris had been offered a new set of boots at the Pole by one of ALE's polar guides, but he had preferred to stick with his own. Oh, how we rued that decision over the coming days.

Fortunately, and by sheer chance, a friend I had met after completing the Scott–Amundsen race back in 2011, Mark Wood, had given me a couple of custom-made brackets that could be used as replacement toe bars. He'd got them from some Norwegians he'd encountered while skiing solo to the South Pole. I suspect he gave them away because he realized they would never be used and were just extra and unnecessary weight. I too doubted that I would ever use them, but both toe bars had remained in my spares bag ever since. With his usual ingenuity

Alex managed a pretty quick repair, which would last the rest of the day and meant we could carry on.

At least Chris seemed to finally understand the importance of skiing correctly, and he managed to stay with the rest of the team for the remainder of the day. It was, however, a short-lived improvement.

In the tent that night I could see that Ollie was also in a pretty bad way. The extreme cold weather we had endured for most of the day seemed almost to have attacked his frostbitten face. His mouth was full of ulcers, and eating and drinking had become a real ordeal. Spicy food in particular was causing him a lot of problems, and so I agreed to exchange some meals with him to help him out. The problem was we had removed all the food packaging to save weight, and I was never quite sure which freeze-dried meal was which. Thinking I was giving him something bland, I inadvertently gave him a Thai green curry dish. As he started to eat it I could see the tears rolling down his cheeks as the ulcers went into overdrive. In typical regimental black humour style, rather than offer sympathy I decided the priority was to capture the whole episode on video. I still chuckle watching it back today.

I was worried about both Ollie and Chris, and whether they would be able to safely complete the expedition. There was a real possibility that Ollie's frostbite could lead to a serious infection, and in the Antarctic that could be life-threatening. I knew with absolute certainty that neither Ollie nor Chris would give in – it just wasn't in their character – but neither would they want to put themselves in a position where they could jeopardize the whole expedition. Listening to Ollie quietly moan in agony that night, I wondered whether he would make it to the finish.

We woke to a layer of deep, soft snow and our progress was slow and hard all day. Despite the effort required, we managed to stay together as a team and, although we finished the day exhausted, our morale was high.

On the morning of 31 December, New Year's Eve, and just when I was starting to believe that we were moving quickly and efficiently, Chris managed to snap the replacement toe bar. It was difficult to know whether to laugh or cry. For a minute or so no one said a word as the full magnitude of what had happened began to sink in.

'I'm sorry, guys,' were the only words Chris could muster.

'Don't worry about it,' Alex said and, proactive as ever, began to search through his pulk for his repair kit. 'If we can bodge it for the rest of the day, I'll have a crack at another repair tonight. But for the time being I'll try and use some sort of temporary fix.'

Alex began working away feverishly, and fifteen minutes later had managed a half-decent repair. His skills were extraordinary and were a real bonus for the entire team. It was fitting that his chosen career should be in repairing broken bodies.

But, despite Alex's efforts, Chris's destructive skiing style made short work of the repair, and within an hour he had removed his skis and walked for the rest of the day. It would have been a disaster for just about anyone else, but Chris's strength and his now recovered piston-like legs probably saved his place on the expedition.

That night, after dinner, I went into Chris's tent to check on progress with the boot and to confront him with the reality of the situation. There was no point in giving him a hard time; he felt bad enough as it was, and he knew that his position on the expedition was at real risk.

As I entered, Alex lifted up one of Chris's boots: 'Good as new.'

'Fantastic work, Alex,' I said with genuine amazement. 'How you doing, Chris?'

'Not too bad, Lou. I know you've come in here to speak to me about my technique and I know I'm now drinking in the last-chance saloon.'

'I'm not going to get on your case, but I've got to make you aware of some facts. The snow is only going to get softer as we descend so walking is not an option. I've informed ALE of the boot issue. They have spares, but we are too far out to be resupplied by Ski-Doo so any resupply will require an aircraft to fly in, and that is going to cost in the region of £10,000 – which we haven't got. Plus, it will mean that the unsupported element to this phase will be over. That is the situation of where we are at.'

Chris's face dropped. 'I'm really messing this up for everyone.'

'No, you're not. We aren't there yet, but that is the situation. All it requires from you is to improve your technique, keep your focus and we'll get through it.'

I spent the next hour or so in the tent just chewing the fat, telling a few war stories and making sure that when I left Chris was in a positive frame of mind. By the time I got back into my tent Ollie was asleep, which was good. He needed all the rest he could get.

The following day, Chris skied like a demon and Alex's repair held. There were no further delays for the rest of that day, and we all felt relieved that – when we reached the campsite that evening – we finished as a team. It was also significant that we had crested the Titan Dome and we were now skiing downhill, even though the decline was barely noticeable.

Chris damaged his toe plate several more times over the next few days, and was forced to walk, but Alex managed to effect more repairs in the evenings. The good news was that Ollie's face seemed to be stabilizing, or at least wasn't getting any worse.

Meanwhile the temperature continued to drop, and eventually reached a low of -40°C. At such low temperatures, any exposed skin will freeze in seconds. Sweat or moisture will quickly turn to ice. Removing gloves at -40°C can lead to an immediate cold injury and the threat from hypothermia increases significantly.

The whole area of the Titan Dome must be one of the most inhospitable places on earth. It was a windswept and desolate mass of snow and ice, and I found myself thinking of Henry, who would have been tackling this same terrain alone and utterly exhausted. We found it hard enough working as a team but, on your own, as Henry had been, almost defies belief.

Fifteen days after we left the South Pole – on 11 January – we reached Henry's final campsite at around 11 a.m. at South 86 degrees 22 minutes, West 176 degrees 00 minutes. It was a desolate patch of nondescript snow, and where Henry had come to the realization that he was unable to continue. It must have been a terrible moment for him.

'This is where Henry's trip ground to a halt,' I told the team. 'He spent three days here being battered by the weather, probably in quite a lot of pain, before being picked up.'

No one spoke. I think at that moment we all realized how utterly demoralized Henry must have felt, despite having achieved and endured so much. But for Henry that would have meant nothing, and he would have accepted defeat with his usual stoicism, knowing that he was one of a long line of distinguished adventurers to be defeated by the Great White Queen.

Prior to leaving the UK, we had decided to honour Henry's legacy. We planned to build a small temporary cairn and hold a memorial service when we reached the top of the Shackleton Glacier.

A few miles later we made camp, taking some comfort from the fact that the worst of the Titan Dome was behind us. It had been another long, tiring day, and passing Henry's final campsite had left us all in a reflective mood. There was little chat or banter that evening and we all seemed content to have an early night.

In the morning we were greeted with a vista that lifted our spirits – the snow-covered peaks of the Transantarctic Mountains; a vast, seemingly impenetrable barrier that almost bisects the entire continent. Literally hundreds of huge glaciers run down through the mountains, and we were initially heading for the Zaneveld Glacier. That would feed us into the mighty Shackleton Glacier – a huge ice highway that would lead us down to the Ross Ice Shelf and eventually our finishing point.

Seeing the distant peaks for the first time was an emotional moment, almost as emotional as seeing the Pole, and it suddenly dawned on me that – had Henry managed to continue for a few more miles – he too would have seen the same peaks beckoning in the distance. I couldn't help but wonder what effect that would have had on him. Would it have provided the impetus to drive him on further? It seemed tempting to assume it would, but tragically the nature of his illness meant further progress was impossible.

The mountains grew in size as the day went on, providing a huge morale booster. The whole team was really buzzing – it was a 'tears in goggles' moment, and we all recognized that we were well and truly on the home run. We had covered around 1,000 miles by this point, with just 100 miles to go. But the terrain was about to change from the flat ice of the plateau to the dramatic terrain of the Transantarctic Mountains. As far as we were aware, there was very little data on the route down through the Shackleton Glacier; all we had were a few waypoints from a team that had come up several years before. There was no guide book, no description of areas to avoid. Nothing.

Initially, the going was easy as we came down off the polar plateau and headed into the mountains. It was the first bit of downhill we had experienced since the start of the expedition. The terrain allowed us to ski down abreast of each other, rather than in single file. There were huge rolling banks of snow which carried us down for some twenty minutes or so before flattening out and starting again. Everyone whooped with joy as we skied down, at times careering into one another or falling over. It was every man for himself. We had a great laugh; for the first time probably in seven weeks, the team could relax and have a bit of fun. I stayed at the back watching the carnage unfold. Skiing downhill with a pulk is not a skill any of us had really mastered. The pulk remained tethered to our waists and, depending on the steepness of the terrain, it could be positioned to the front, side or rear. In theory the idea was simple, but in practice the pulk took on a life of its own and often behaved like an out-of-control dog, pulling us one way and then the next.

The Titan Dome area was finally cleared on Day 18, 14 January, and we skied into the gateway to the Zaneveld Glacier, an area dominated by a large, snow-free rocky outcrop called Roberts Massif. This covered an area of around ten square miles, rising to over 8,800 feet. Everything in Antarctica is on an epic scale. The plan was to stop for a day or so and conduct the memorial service for Henry. For me, this was a poignant moment in the expedition. Henry was my mentor. If it had not been for him, I would have never visited Antarctica, so I felt a huge sense of responsibility that a memorial should take place on the continent he loved so much. But there were some significant issues.

I had wanted to build a cairn as a lasting tribute to Henry and to serve as part of his legacy. Having witnessed first-hand, back in 2011, the wonderful impact that finding Amundsen's cairn had had on him, it seemed most appropriate. In my eyes he was one of the great Antarctic explorers and deserved some sort of official recognition. But prior to leaving the UK, I had been informed by the Polar Desk at the Foreign Office and later by ALE that building any memorial – even one out of a few stones – would breach the rules that prevent any form of alteration, no matter how small, to the landscape. I had a very good relationship with ALE and I didn't want to impact it adversely.

I *was* allowed to build a cairn providing that, after the service, the structure was dismantled and the rocks replaced in their original positions.

By the time the team arrived at the outcrop and found a suitable location for our camp, it was already late afternoon. While the rest of the team began erecting tents, James and I gathered together some climbing gear and went to try and find a suitable location to hold the service.

I looked out across to the main area of Roberts Massif and

mentally plotted a route across an area of uneven ice. By now my Antarctic instincts were finally honed and, although I couldn't see any obvious problems, I assumed there would be a few crevasses as the ice sheet butted up against the mountains.

I turned to James and, pointing in the direction of Roberts Massif, I said: 'If we head across the ice using the main part of the massif as our aiming mark, we should be OK. Ground looks a bit disturbed so tread carefully.'

'Yep – you lead, I'll follow,' James added.

I stepped off carefully into the deep, soft snow, adding weight to each step and almost trying to feel the thickness of the snow with my feet. At the back of my mind something was telling me that we should be roped together, but we were keen to make progress and we pressed on regardless.

After about thirty minutes, and just as I was beginning to feel more confident about the route, the snow gave way and I fell into a crevasse. For a second, as I dropped, I thought I was done for. I instinctively reached out, digging my hands into the snow, my arms and shoulders on the surface while the rest of my body dangled over the edge. I carefully moved my legs to see if I could grip the icy walls of the crevasse, but it was no good. I was effectively hanging above a vast chasm.

If I moved too quickly the snow bridge could give way even more and I would plummet to my death. I breathed slowly and forced panic from my mind. I turned, looked at James, and smiled.

'Hang on, Lou,' said James, who was standing about five yards away.

'Slowly, James,' I cautioned. 'No rush. I'm going nowhere. Watch your footing, otherwise you'll be going in as well.'

James inched forward, grabbed my shoulders and carefully hauled me out. My heart was pounding like a racehorse as I turned onto my knees and counted my blessings. I took a few deep breaths and waited for the adrenaline to bubble away before peering into the bottomless void.

'We need to take real care,' I said to James. 'I think this place is riddled with crevasses.'

A few steps further on and it was James's turn to fall into a crevasse, just as I had done a few moments earlier. He too was hanging on with his arms. It was roles reversed and I hauled him to safety.

It is an unnerving experience, and a rare moment in your life when you examine your own mortality. It is the equivalent, I would say, to walking through a minefield.

Eventually we managed to get onto solid ground on Roberts Massif and, as we climbed to the top, we were rewarded with a stunning panoramic view back across the polar plateau and north down the Shackleton Glacier – a five-mile-wide, 100-mile-long highway of pure blue ice leading down towards the Ross Ice Shelf in the far distance. It was almost as if the view was the physical embodiment of Henry's spirit.

'This is the place,' I said to James, who nodded and initially seemed lost for words as we took in the raw natural beauty.

'Yeah,' said James a few seconds later. 'There is no better place than this. It is simply awe-inspiring.'

Not for the first time I could almost feel the presence of Antarctica as I gazed across the landscape. It was a rare moment where the continent seemed to be rewarding us for our efforts, and I knew deep down that Henry would have approved.

'James, you head back down and meet the others and I'll start building the cairn. Take care and make sure they are roped together when they cross the crevasses.'

As James set off, following in our tracks but taking care to avoid the crevasses, I began to gather stones from nearby and build the base of the cairn, modelling it as closely as I could on the photographs Henry had showed me of Amundsen's cairn. The images were burned into my memory.

An hour or so later, James returned with the rest of the guys, and we finished the cairn together. Two hours later we had managed to assemble a fairly substantial five-foot-high structure and we were all pleased with our efforts. I had carried Henry's compass – the one he had used on his last trip – and I placed that on top of the cairn. We all gathered round and I pulled out a piece of paper from the inside of my jacket.

I had thought long and hard about what to say at this moment, and decided the most fitting words were those from a speech given by Theodore Roosevelt, who served as the United States President from 1901 to 1909. His words described what Henry was and what he was trying to achieve.

And so I began: 'It is not the critic who counts; not the man who points out how the strong man stumbles, or where the doer of deeds could have done them better. The credit belongs to the man who is actually in the arena, whose face is marred by dust and sweat and blood; who strives valiantly; who errs, who comes short again and again, because there is no effort without error and shortcoming; but who does actually strive to do the deeds; who knows great enthusiasms, the great devotions; who spends himself in a worthy cause; who at the best knows in the end the triumph of high achievement, and who at the worst, if he fails, at least fails while daring greatly, so that his place shall never be with those cold and timid souls who neither know victory nor defeat.'

A minute's silence followed as we reflected on Henry's journey. Then I spoke about the nature of the man and how much of an inspiration he was to those who knew him. Henry was a real fan of cigars, so we all smoked one and talked about how far we had come over the last seven weeks and how we had all been changed by Antarctica.

In that moment, as the other guys chatted amongst themselves, I became completely aware, for the first time, of how much of a friend I had lost. The pace of my life had been so frenetic over the past twelve months that I had barely had time to comprehend or make sense of Henry's death. But in that moment, even though I was surrounded by a group of great individuals and good friends, I felt adrift. I suppose it was the way someone feels when they lose a brother.

With the service over, I removed Henry's compass and we began to dismantle the cairn. It was late by the time we returned to our tents, and overnight a storm moved in and hit us with 40 mph winds that produced a wind chill of -45°C. The storm continued throughout the next day and we remained tent-bound for the whole of 15 January. The rest day was welcome, and Ollie and I watched the movie *Interstellar* on my iPhone while the others slept or played cards. Later that evening, James, Alex and Chris braved the outdoors for a few seconds before charging into our tent for a party in the Ritz. We played Cards Against Humanity, where one player asks a question from a black card – such as what was the worst thing we'd done at school – and the other players have to use one of the white answer cards they've drawn to respond in the funniest way. But, as the saying goes, 'What happens in Antarctica stays in Antarctica.'

Although there were still around a hundred miles left to cover, it was all largely downhill, and there was a real feeling amongst the team that we were on the home run. I knew that these were precious days, which I would soon miss once the expedition was over. Whilst I was looking forward to getting home and being reunited with my family, the sense of camaraderie I had experienced in Antarctica was unique. We were a group of individuals bonded into a tight-knit team by a common goal and shared adversity.

As is so often the case in Antarctica, the violent storm that had kept us locked inside our tents had disappeared without any trace by morning, leaving in its wake a perfectly blue, cloudless sky. So by 8.50 a.m. on the morning of 16 January, we were on the move with me leading. The sun shone brightly and the omens seemed good. Morale in the team was high and there was something of a collective relief that we had managed to conduct Henry's memorial in such a dignified manner. Up until that point, Henry's presence had been with us all the way, but before departing for the final push to the finish point, I told the team that the remainder of the expedition belonged to us. We set off that morning fully intending to enjoy the final few days.

The plan for that day was to traverse around the edge of the Roberts Massif, which should have led us into the Zaneveld Glacier, a feeder glacier to the Shackleton, and then on to the main glacier itself. It was a tough, slow haul due to the deep, soft snow. At the same time the weather began to deteriorate, until a couple of hours later we found ourselves stumbling along in complete whiteout. We were hugging the edge of the Roberts Massif on our left, and inadvertently followed it round into a kind of cul-de-sac that led us into the heart of the feature. We stopped and did a navigation check and quickly worked out what we had done.

'Right, guys, I've messed up,' I told the team, who were more surprised than annoyed. 'Basically, we are going the wrong way, so we have two options. Backtrack from where we came, but that is going to take at least an hour. Or we can go over that ridge to our front and see if we can drop down onto the Zaneveld. Might be a bit steep though.'

The shortcut was the unanimous preferred option, but we soon realized that we had probably bitten off more than we could chew. The climb up to the ridge through the soft snow was ridiculously difficult, and the other side turned out to be incredibly steep and tricky. Over the next two hours we picked our way through boulders, hauled each other out of deep snow pockets and lowered the pulks down on ropes where the terrain was simply too steep. Steps were kicked out of the ice to help with our descent. The pace was agonizingly slow, and soon the wind speed began to pick up; very quickly we all began to feel the effects of the wind-chill factor in our exposed position. Within a matter of minutes our face masks began to freeze, our body temperatures plunged, and we soon began shivering in the cold, icy wind.

I began to have serious concerns about the route, but it was now too late. Every disaster begins with a small misjudgement, and events can quickly snowball. But we had reached the point of no return. It would have been far harder and very dangerous to try to re-climb the ridge. With hindsight we should have turned around, taken the pain and followed the original route back. But such is the life of adventurers, or so I told the team. Secretly, however, I was quite worried about someone sustaining an injury or damaging some gear. A medical evacuation would have been very challenging from this position.

Lady Luck was on our side, however, and fortunately we all made it down in one piece – but an important lesson had been learnt. I mentally noted that next time we came across a route problem, we should go for the safest and – if necessary – the longer option.

It had taken three hours of hard work, but once we were on the Zaneveld Glacier we flew down and the going was super-easy, almost like skiing on a piste – but it wasn't to last. By the time we arrived at the end of this small feeder glacier, the snow had started to disappear. The Shackleton Glacier was exposed to the wind; it seemed that as soon as any snow fell it was blasted away, leaving only a diamond-hard surface of blue glacier ice.

Our camp that night was made on the last remaining piece of snow in the area. The tents were huddled together, almost touching so as to avoid pitching on the hard, rippled ice. The limited amount of snow meant that the only way we were going to get enough water to rehydrate our meals was to laboriously chip away at the literally rock-hard ice for at least an hour.

Over the next forty-eight hours the terrain changed markedly. The glacial ice was smooth, as if a gentle river – working its way peacefully down through the mountains – had suddenly frozen, and although beautiful it posed a major hurdle. Any attempt to ski on glacial ice would have been carnage because of the lack of grip, and so we swapped our skis for crampons. The whole of your body weight is then placed on a few spikes, which dig into the ice for grip. Up to this point we had been sliding on skis for two months and 1,000 miles, so returning to a more normal walking motion caused us a few problems with aching joints and blisters. On the positive side, though, the pulks could be pulled along with almost minimal effort – so it was swings and roundabouts.

Within a day the terrain had changed again and the glacial ice appeared to be riddled with crevasses, some just a few inches wide and a couple of feet deep, but they made movement very difficult and slow. Once again, it was almost like walking through a minefield, knowing that one step in the wrong place could end in disaster.

The air was filled with curses and swearing as everyone was slipping, jarring knees and ankles or being pulled backwards by their pulks as they got stuck in the ice. Our movement as a team slowed to a crawl, and I was constantly sending guys out to see if they could find routes to the left or the right side of the glacier. This went on for mile after mile and hour after hour, and was a further reminder that our journey in Antarctica was far from over. The only positive was that we were going downhill and the temperature was improving every day, which was good news for Ollie, who was still suffering with his cold injuries.

The end of the third day of the descent, 18 January, almost ended in disaster, for me at least. We were looking for somewhere to erect the tents – ideally a piece of undisturbed snow or some flat bare rock; in fact, almost anything but ice. After twenty minutes scouting the area, Chris spotted a suitable campsite, and we all quickly got to work as the cold began to slow our movements. As soon as Ollie and I got our tent up, he began to move our sleeping bags, food and cooker inside, while I adjusted the guy lines to ensure it was stable enough to withstand anything the evening might throw at us. Just as I moved around the back of the tent, my left leg disappeared into a hidden crevasse. The whole of my body weight fell forward, bending my knee the wrong way so that it was hyperextended. It felt as if I had been hit by a high-velocity bullet and I collapsed in a heap. The pain was so intense I was convinced that my leg had been broken. Ollie, who had seen me fall, rushed around to where I had slumped on the ice.

'Lou's down!' he shouted. 'I need some help here.'

He must have seen that I was in incredible pain because there was no piss-taking.

'Just hang on, Lou, and we'll pull you out.'

Alex, Chris and James appeared within seconds, and slowly began to extract me from the crevasse. I was waiting for the pain to subside as the adrenaline kicked in but instead it was getting worse. My mind quickly went into overdrive and I began planning my evacuation. I was breathing heavily, feeling nauseous and sweating, even though it was -20°C. Lying on my back, I was almost unable to move.

'Bloody hell, Lou,' Chris said as he stared into the crevasse, 'I can't see the bottom of that one. It must be pretty deep. Count yourself lucky.'

Alex rolled up the trouser leg of my insulated salopettes and gently began to examine my leg.

'Well, it's not broken,' he announced with undisguised relief. 'But it's going to swell up and hurt for a while. Try and stay off it tonight and see how you feel tomorrow.'

'Painkillers, please,' were about the only words I could muster. Chris made an ice pack and rested it on my knee, and I decided I wasn't going to move until the pain subsided. Eventually, after half an hour or so, I managed to crawl inside my tent, where on further examination Alex and Ollie agreed that I had probably strained some ligaments and torn the cartilage in my knee. Over the next few hours the pain eased and my knee eventually appeared to stop swelling. It was a huge relief when I realized that my expedition hadn't come to a premature end. The next few days, I told myself, would be tough but achievable.

That night I chatted to the doctor at ALE and underplayed the pain I was in, merely explaining that I had jarred my knee and had taken painkillers.

After a painful night of little sleep, I woke to find that the swelling had gone down but my knee still ached. I carefully got dressed under Ollie's watchful eye and, now with the tables turned and me being the one suffering, he seized his opportunity to relentlessly take the piss for the next couple of hours. I took all the banter on the chin and laughed at Ollie's terrible jokes. I had my breakfast and then took as many painkillers as I could get away with. After breakfast, I virtually slid out of my tent and Ollie helped me stand.

'We are going to empty out most of the kit in your pulk, Lou, so you are pulling next to nothing. No arguments,' Ollie said. 'Then we'll assess at lunchtime and see how we are doing. The first part of the day will be tricky. It's ice, so we'll be on crampons, and the last thing you want is to make a bad situation even worse.'

'Sounds like a plan,' I responded. I wasn't going to argue about not pulling my weight in these circumstances.

Trying to walk down an ice glacier on crampons with a severely dodgy knee is not advised, and I was the epitome of the elephant on roller skates.

As the hours passed my knee seemed to settle and the pain eased off; when it became clear that I was going to be OK, the banter began again. I was the old man, the casualty, sick-note and the lightweight. Fortunately, in the afternoon the terrain improved dramatically and became flatter and smoother, making walking easier – being dosed up to the eyeballs on painkillers also helped.

The following day, twenty-three days after leaving the South Pole, we reached the mouth of the glacier and the area flattened out into a moon-like, desolate landscape dotted with spectacular and almost surreal ice formations that seemed to have been pushed out of the ice by some unknown force. It was like nowhere else we had experienced since our arrival in Antarctica. My knee had improved considerably, and I was only reminded of the previous day's damage by a small jolt of pain every time I stepped on slightly uneven ground.

The final few miles of the glacier looked like a frozen lake and were devoid of crevasses, which enabled us to walk easily. We were now rapidly approaching sea level and the temperature change was remarkable. It was still -10°C but felt like a summer's day in comparison to the temperatures we had experienced on top of the Titan Dome.

After a mile or so, the frozen surface we were travelling along narrowed and we ended up walking alongside a narrow channel around a 20-yard width of quite weak, thin ice with running water below. Gradually we realized we were getting pushed off bearing and would need to cross at some point.

'Any ideas?' I said to the guys.

The responses were muted, possibly because it had been a long day.

'Well, we could slide over on top of our pulks,' I suggested. 'If we try and walk across we'll probably go through the ice, but the pulk should help disperse the weight.'

I received dubious looks from the rest of the team whom I sensed doubted my plan.

'OK, I'll go first,' I said confidently.

I positioned my pulk and took a few steps back. Powering off with my good leg, I dived onto my pulk and slid across effortlessly.

'Piece of cake, who's next?'

Ollie, James and Alex quickly followed, leaving Chris.

'Chris, you are really going to have to take a long run-up, mate!' I shouted, suddenly appreciating that Chris was heavier than the rest of us and the ice might not take his weight.

Chris dived onto his pulk and flew across, but almost immediately the ice began to give way and his progress slowed. Then there was an almighty crack beneath his pulk and the ice began to shatter and collapse behind him.

Fortunately, his momentum had carried him far enough across, and we were able to grab him before he slid backwards and into the deep, icy crevasse below. In isolation it might have been a pretty scary experience for Chris, but set in the context of a trek across the continent, it was greeted with uncontrolled, raucous laughter.

My diary evening for that evening read:

Amazing day. Terrain improved dramatically. Initially, we made good progress despite my knee. Managed to get loads of great downhill toboggan runs on the pulk. Continued until 2000 hrs and managed to achieve 19.1 nm. Weather still beautiful and only 12 nm from pick-up. Surface gradually changing to snow now.

The final full day of trekking began with mixed emotions. We set off at 8.30 a.m., spurred on by the excitement of reaching our goal, but I also sensed that we all felt something very special was coming to an end. The Ross Ice Shelf was now visible just a few miles ahead of us – a vast, flat white area of ice 10,000 feet thick that seemed to go on until it reached the horizon many miles away. It is in fact the largest floating ice shelf in the world and covers an area roughly the size of France.

At around 2 p.m. we cleared the grounding line, which is the point at which the glacier starts to float. Skiing line abreast for the first time since our arrival at the Pole, we skied off the continent and onto the Ross Ice Shelf. We all cheered and whooped with joy – we had done it. Everyone was buzzing, and we briefly stopped and stared back at the continent and the mountains behind us. It was a truly momentous moment for us all.

Despite arriving at the Ross Ice Shelf, we decided to carry on skiing for another five miles, just to be 100 per cent certain that we were well and truly off the continent and clear of the high-velocity ice at the mouth of the glacier. Besides, we didn't want our claim of a traverse questioned at any point in the future. As I was later to find out, the polar community can be a very judgemental place.

The campsite location that evening was an arbitrary position out on the Ross Ice Shelf, chosen because it was very close to a stretch of ice that appeared flat enough to allow ALE's Twin Otter aircraft to land. Later that evening I contacted ALE, told them that we had reached the finish point and required extraction. I gave them the coordinates of the proposed landing site. These were subsequently fed into a computer software program that is equipped with radar mapping and specifically designed to be able to reveal faults in the ice.

A few hours later, the ALE team contacted us by satphone and said: 'Hi, guys. Thanks very much for your proposed landing site. Now we appreciate that you can't see this, but our radar imagery says that the area you've chosen is an area of high-velocity ice. We'd like you to move 2.8 nautical miles to another set of co-ordinates we are going to send you which will provide a safer landing for the aircraft.'

I stared at the phone slightly confused. The area we had chosen looked perfect, but I was too tired to argue.

'Roger,' I responded with a sigh. 'I'll call again when we get to the new pick-up point.'

To be honest it was a bit frustrating, especially as we had pitched our tents and marked out the landing site with our pulks. But we cracked on, nonetheless. The camp was dismantled and we were soon on our way again.

When we arrived at the exact latitude and longitude they had sent us, we simply could not believe it. The proposed pick-up position was directly on top of a massive crevasse, deep enough to swallow a whole aircraft. As we scouted around the area, Alex fell into a crevasse up to his waist while still wearing his skis and had to be pulled out by other team members.

I called ALE again: 'Guys, we are at the exact location you asked us to move to and it is on top of a huge crevasse – the area is riddled with them. I think it's completely unsuitable for a PUP.'

'OK. Accept that. Try and find a suitable area away from the crevasse. Thanks.'

After a couple of hours scouting around, we managed to find an area of ice that met everyone's requirements, although the proposed airstrip was only just long enough. The pulks with their bright red covers were positioned along the sides of the airstrip so they were visible from the air, but as we sat and waited, the satphone rang again to tell us that bad weather at their end was going to delay our extraction by another twenty-four hours. We had a relaxing day and another party in the Ritz, where we finished the remaining booze and chatted about the highs and lows of the expedition. My thoughts turned to Al and I wondered how he was getting on.

The following morning, we waited with barely controlled excitement for the aircraft's arrival. Alex was the first to hear the low, distant hum of the engine as it approached the makeshift landing zone.

'There it is,' he shouted, pointing at a barely visible distant dot high in the sky above the Transantarctic Mountains.

Then we all saw the aircraft. It came in high and slow. Initially it headed over to our first proposed landing site and circled above it for a while before heading back to us and coming in low to check out our makeshift runway. Once they were happy, they set themselves up on the approach and put it down right between our pulks, which gave me a real sense of satisfaction that we had picked a good spot. I had marked many a Tactical Landing Zone (TLZ) for military aircraft in the past, but never onto ice before.

We had deliberately positioned ourselves away from the landing strip, close to the edge of the crevasse that Alex had fallen into earlier, so as to ensure the aircraft was kept well clear of it. But in a final moment of comedy we watched as the aircraft began to taxi back along the strip towards us. We frantically waved to try and warn them, but they obviously took our gestures as some form of excited greeting and taxied right up alongside us and parallel to the crevasse.

When the aircraft eventually shut down, the pilots emerged, greeted us all with a hug and then confessed that our original suggestion for an ice-landing strip was probably better, especially when we pointed out the nearby crevasse. But we cared not. Photos were taken, we all hugged each other again. There were a few tears and then we climbed aboard for the six-hour flight back across the other side of the continent to Union Glacier. The flight was cramped and uncomfortable and the noise of the engines made conversation difficult, so I looked out over Antarctica, feeling amazed at what we had achieved. We hadn't conquered the Great White Queen, no one ever will, but she had allowed us to achieve our objectives, and for that we were grateful. As the minutes ticked by, the weeks of endless fatigue kicked in and pretty soon all five of us fell into a deep sleep.

The Twin Otter touched down around 10 p.m., and we received a heroes' welcome from all the ALE staff and other expeditioners at UG. As we exited the aircraft, we were plied with champagne, cheered and applauded by a large reception committee, which also included some of the ALE directors who had flown in from Chile for the celebration. It was all completely unexpected, and I think we hadn't fully appreciated what we had done until someone said, 'How does it feel to be the first British team to cross Antarctica?'

A special meal had been laid on and we stayed up until 4 a.m. drinking and telling stories to anyone with the staying power to listen.

The next morning at breakfast, as we reminisced about the night before, we were approached by a gentleman named Richard who was the organizer of the Antarctic Ice Marathon. He congratulated us on our achievement and then asked whether we would be interested in taking part in the marathon due to take place the following morning. The race was part of the Seven Continents Marathon Club, an event where athletes run seven marathons in seven days on the seven continents. I laughed, thinking it was a joke, but Alex, Ollie, Chris and James looked at each other and said: 'Why not?'

Alex, who was particularly keen, added: 'We will never get another chance like this in our lives. Let's do it.'

I was lost for words. We had just skied 1,100 miles across Antarctica and now these lunatics were planning to run a marathon. The four of them then set about scrounging kit, including running shoes, and twenty-four hours later they were on the start line with a large group of international marathon runners who had flown in especially for the event. I wasn't convinced that any of my team would complete the course, which took them on a series of loops around a cut track at UG, but they all did and incredibly Alex, competing against an international field of more than forty other runners, came fourth! The other competitors were devastated that they had been beaten by someone who was recovering from having just completed a crossing of Antarctica.

We finally left Antarctica on 26 January when we flew out with the rest of the ALE staff. The short expedition season had come to an end and winter was looming. With regular storms, perpetual darkness and temperatures reaching as low as -80°C, Antarctica is not the place to be outside of the summer months.

We spent three days in Punta Arenas sorting out our equipment prior to it being airfreighted back to the UK. Then we flew up to Santiago, where the team was very kindly hosted by the British ambassador and her husband at their personal residence. It was a wonderful few days, and I gave a small presentation on the expedition to some gathered local dignitaries at an embassy cocktail party. It was a taste of things to come.

Post-expedition blues began to kick in on the flight home as I began to contemplate my immediate future. I had experienced similar mixed emotions before going to and returning from operations with the Regiment. Before setting off you often feel guilty about leaving your family behind, not knowing if you will ever see them again and wondering how they might cope with your loss. Then when you arrive in the theatre of operations, you enter another world. In all honestly, your other life is almost forgotten, or at least parked in a place where it can't interfere with your operational life. Then, when it's all over and you arrive home, you feel almost lost, detached, and wondering what your reality is – the war zone or your home life. It can take days to adjust to the transition and, as much as you love your family, you have a selfish longing to get back into the action.

The SPEAR expedition had been an all-consuming part of my life for the previous two years, and now suddenly it was all over. I was due to leave the Army in about three weeks. I swallowed hard and shuddered. Leading SPEAR had been an amazing opportunity, but I now had to think seriously about my future and how I was going to pay the bills.

13

ANTARCTICA'S LITTLE SISTER

By failing to prepare, you are preparing to fail.

BENJAMIN FRANKLIN

By the end of March 2017, my tenure as RSM had come to an end, and with it my twenty-five years of service in the SAS. But my professional life had suddenly become so busy that I had little time to reminisce; and besides, you never really leave the Regiment – it's always a part of you. In the weeks before the start of SPEAR my future seemed uncertain, but I learnt very quickly after returning from Antarctica that I had been promoted to the rank of captain. Prior to setting off on SPEAR, I had applied for what is known as a Late Entry Commission, in the Parachute Regiment, and discovered on my return that my application had been successful. I was informed that I would be taking over a post at the Infantry Battle School in Brecon, South Wales. It was an important – if not particularly glamorous – job. The Infantry Battle School is responsible for training young infantry officers and non-commissioned officers on their tactical battlefield roles. My job was to provide support for all of the numerous courses taking place at any one time.

The job had been gapped for six months – meaning my post hadn't been filled – so it required a lot of work just to get it up and running efficiently again. Fortunately, I had a great team of around forty-five staff, composed of civil servants and service personnel, to help.

The months ticked by and I got stuck in. Memories of Antarctica were quickly buried amid an ever-increasing workload. Although my thoughts returned to the continent periodically, when I did think about it, I didn't see a way I could return. The next best thing was to be involved in someone else's Antarctic plans.

The year after the SPEAR expedition, an all-female Army team completed a similar crossing of Antarctica, titled Exercise Ice Maiden. They benefited massively from our experiences and I was involved in advising the team on many aspects of their trip. It ended up almost being a carbon copy of the SPEAR project: there were also six of them and they were able to use our original pulks and tents, which we had returned to the Army. Using the contacts I passed on to them, they ended up in exactly the same blue polar suits from Mountain Equipment, the same skis, poles and boots, sleeping bags and communications equipment. In photos of the team on expedition, with their hoods up it was hard to tell the difference between them and SPEAR. The Army asked me to sit on the panel that was part of the approval process for their expedition and it was clear they had prepared well and were taking a sensible approach to resupplies along the route. Ultimately they had a great expedition and successfully completed the crossing. On their return, they claimed three world records: first female team, largest team and first polar novices. I think you have to be quite careful doing this as these records are purely self-proclaimed. There's no world governing body that has bestowed these accolades, no certificate or piece of paper with official acknowledgement. And where do you draw the line in claiming distinctions for world records between your trip and others that have gone before? I think SPEAR should have claimed the world record for the most booze carried on a polar expedition!

In November 2017, the adventurer Ben Saunders began an unsupported, solo traverse attempt. Ben was a seasoned polar

Above: An incredible piece of silverware made to commemorate our expedition, which will be part of the Regiment's collection.

Below: Leading a team of civilians on a crossing of Greenland presented different challenges. *Top row, from left:* Chris Brooke, James Redden, Pete Masters and me; *bottom row, from left:* Arabella Slinger and Wendy Searle.

Above: A chance encounter with Norwegian polar guide Bengt Rotmo, who was also leading a team.

Below: In the middle of the Greenland Ice Sheet in poor visibility.

Above: Pete and I re-enacting the cigar photo with Henry at the end of the Greenland crossing.

Below: Collecting my MBE for leadership on the SPEAR17 expedition, a week before I departed for Antarctica again for my solo crossing.

Above: Final training and equipment testing on the Langjökull Glacier, Iceland.

Below: Everything I would need for the solo crossing, apart from 75 days of food and 19 litres of cooker fuel.

Above: Surely the coolest boarding pass in the world for the flight to Antarctica.

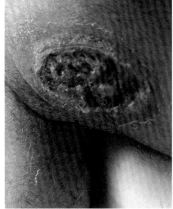

Left: Ice tusks created purely by the moisture from my breath freezing, as I struggled on across the polar plateau towards the Transantarctic Mountains.

Above: An open polar thigh wound that started from a small abrasion injury. A love bite from the Great White Queen.

Above: Reaching the South Pole for the third time, but this time alone and with no stopover or resupply.

Below: A wonderful message of support from the South Pole station team. I only stayed for around thirty minutes for some photos before heading back out into the wilderness.

Left: My 12-kilogram weight loss – but I felt good for it. If I'd been given a resupply I could have gone on if needed.

Right: The ALE pilots very kindly brought me a bottle of champers for the flight home. Went straight to my head in the unpressurized cabin.

Above left: A huge privilege to be able to run part of the thirty-miler across Dartmoor with my son Luke and present him with his green beret at the end.

Above right: A proud day watching my daughter Sophie being called to the Bar as a barrister.

Below: Another one joins the mob! With Amy joining the RAF, we ended up with three of the family serving at the same time.

adventurer, an endurance athlete, younger and a little more experienced than Henry, and I fully expected him to succeed.

Roughly around the time Ben was making his way across Antarctica, I decided to plan and lead a civilian expedition on a month-long, 350-mile trek across the Greenland Ice Sheet, from the west to the east coast. I'd got the idea after a group of civilian friends – best described as an eclectic bunch – had said that they would like to take part in a polar expedition. I'd wanted to complete a traverse of Greenland for quite a while after reading about the Norwegian Fridtjof Nansen's fabled crossing (it's classed as part of the hat-trick of polar journeys, along with the North and South Poles), and the timing worked. The traverse is best done in the spring, and luckily there would be an Easter break in the programme at Brecon.

The first member of the team was Wendy Searle, who had been the media manager for SPEAR. She had done such a great job and had become so inspired by the project that I promised her I would one day take her on a polar journey so she could experience an expedition from the other side. Next was Pete Masters, who was something of a local legend in Hereford. Pete is one of those larger-than-life characters, and always tells it how he sees it. He had been through a few rough periods in his life, and was now the proud owner of Skinzophrenic Tattoos in Hereford. I came across Pete when I got a tattoo of the map of Antarctica done, after returning from SPEAR.

Pete was and is an intriguing character, probably best described as a lovable rogue. He is one of those people who gives all or nothing. He walked past as one of his artists was etching away and began asking why I was getting it done. By the time I had finished explaining, he was hooked on the idea of a polar adventure. We agreed a progressive pathway to build his experience, which started with some fitness training, and then we headed off to Norway on a two-week polar training course run by a good friend, the highly experienced polar guide Hannah

McKeand. Pete's motivation was in part to become a better role model for his young son, so I took him under my wing, determined to help him achieve his dreams in any way I could. When I told Pete that he was on the Greenland team, he was incredibly emotional and vowed not to disappoint me. He subsequently read a book about Shackleton and how his men called him the 'boss'; so as far as Pete was concerned, I was the 'boss', and from that moment until this day that is how he refers to me. Which, secretly, I kind of like.

Next up was Arabella Slinger, who worked in finance in London. She had learnt through word of mouth in the tight-knit world of polar expeditions that I was planning a trip to Greenland. She was very tenacious, single-minded, and had already completed expeditions in both the Arctic and Antarctic. She had even tackled a polar bear in Svalbard, a Norwegian archipelago between mainland Norway and the North Pole. The bear had ripped open her tent in search of food but instead found Arabella asleep. She woke with a fright as the bear was ready to make her its next meal. Her rifle was out of reach so instead she picked up a cooking pan and belted the bear across its snout before rolling out of the tent. She eventually managed to retrieve her rifle and forced the bear to retreat with a few carefully aimed warning shots.

The fourth member of the team was James Redden, a former soldier who had served in the Royal Signals. James had heard on the grapevine that I was planning a trip to Greenland and asked to come along. He was very fit and highly motivated, and I was happy to have him on the team.

The last member of the group was Chris Brooke from the SPEAR team, who became my second in command. I trusted Chris with my life and knew that he was 100 per cent reliable.

The plan was for us to self-organize rather than use a specialist company, which meant that we, and mostly I, would have to arrange all the flights, the shipping, the accommodation and all the necessary permits required to complete the crossing. It was going to be a lot of work but the cost would be reduced significantly, from £17,000 per person to around £5,000.

I used the same training template for the Traverse of Greenland (TOG) expedition as I had done for SPEAR. We met up on a few weekends and trained, dragging a tyre on a pulk harness across the countryside and running through all the routines. Each member knew what was expected of them and the level of fitness required to successfully complete the crossing. Of all the team members, Pete was the one I worried about most. He is a man with a huge heart, but he doubted whether he had the necessary physical and mental determination required. I knew that taking Pete along was risky, but I was also aware that if he could last the course of the expedition, it would have a positive, life-changing impact on him. I told the team that they would have to do most of the physical preparation themselves. I could give them exercises and advice, but the main effort would be up to them and, I explained, the workload should not be underestimated. Self-discipline was the order of the day.

'No matter how hard you think it is going to be,' I said with a smile, 'it will be harder. You need to train hard, both body and mind, because it will be as much of a mental battle as a physical one.'

In the meantime, I was monitoring Ben Saunders's unsupported solo expedition. Within a matter of weeks it became clear that his progress towards the Pole was being seriously hampered by sastrugi and whiteout. So much so that, by the time he arrived at the South Pole in December 2017, he called it a day. Ben had just fourteen days' worth of rations left, not enough food, by his estimates, to continue. Almost immediately, friends within the polar community started asking whether I was going to attempt the same journey. At that time I had enough on my plate, coping with my still relatively new role at Brecon, and I doubted whether the Army, let alone my family, would be too impressed if I said

I was going back to Antarctica so soon. But the seed was sown, and over the next couple of weeks I found myself working through the numbers. It was doable, I concluded, but only just. An expedition launching in October 2018 would give me ten months' planning time. By that point I would have been in my job for almost eighteen months, so the Army probably wouldn't be too miffed if I told them I was attempting another expedition.

I fired off an email to ALE and asked whether they thought I had the requisite experience to undertake a solo traverse expedition. Their support was crucial; if they had said no, that would have killed off any plan, but to my delight they confirmed they were happy. They also told me that, following the end of Ben's attempt, they had received a whole raft of enquiries from around the world. So far they had a shortlist of around five serious contenders. If I was going to book my slot with ALE, I was going to have to be quick.

That evening I sat down with Lucy and put my cards on the table.

'You know Saunders's attempt at traversing Antarctica failed,' I said, as matter-of-factly as possible.

Lucy was reading on her iPad and acknowledged my question with a 'hmm'.

'Well,' I said, but before I could continue, Lucy put down her iPad and stared at me.

'Well?' she said.

'Well, I thought I might give it a go.'

'Go for it,' she said. 'I know how much this means to you and I'll always support you, you know that.'

I was delighted. Had Lucy said no I would have canned the idea, but the door was now open.

I also contacted Henry Worsley's wife, Joanna, and told her I would only go ahead with the expedition with her blessing. I didn't want her to feel that I was attempting to crack a journey that I still believed should have belonged to Henry.

My main challenge, as with all polar expeditions, was going to be the fundraising. Somehow, I was going to have to find £150,000. ALE was a business, and they weren't going to be offering any discounts. All payments had to be made up front and failure to raise the cash was the main reason why 99 per cent of polar expeditions never get off the ground.

Very quickly I realized I had bitten off more than I could chew – and that was an understatement. My job at Brecon was still demanding, while the planning for the Greenland expedition was becoming increasingly time-consuming and frustrating. I had seriously underestimated the bureaucracy involved in getting the necessary permits from the Greenland government, and it had become a logistical nightmare. I had less and less time to spend planning for the Antarctic crossing, and I often found myself on my computer at 3 a.m., banging out a letter to a prospective sponsor, only to receive a rejection a week or so later.

By February 2018, the possibility of both expeditions grinding to a halt was very real, and so I made the difficult decision to park the fundraising for the Antarctic expedition until I returned from Greenland. It was a good move. The workload involved in trying to simultaneously organize two expeditions while in a demanding full-time job had left me close to physical and mental exhaustion for the first time in my life. Up until that point, the most stressful experience of my life had been my first taste of combat when I was young and relatively inexperienced. Then I had been living life hour by hour, fully immersed in the operational challenge and the demands from senior officers, who seemingly wanted me and my team to achieve the impossible on a daily basis, but by contrast with organizing these two expeditions simultaneously, war seemed like a walk in the park.

Also, by that stage the Greenland team had invested a lot of time and money in the project, and there was no way I was going to let them down. If needs be, I was prepared to sacrifice my

Antarctic solo attempt, but I was still sure back in early 2018 that I could bring it all together within eight months.

In mid-April, roughly two weeks before we were due to depart for Greenland, I arranged for the team to meet up in Birmingham to begin packing the kit, which needed to be dispatched by airfreight. The tents, sledges, skis and all the food had to be broken down into smaller useable bags.

I'd sensed that Pete was starting to get a little bit anxious about the size of the challenge and, on the day we were due to meet, he called Wendy and told her that he wasn't feeling very well.

As soon as Wendy explained what had happened I got on the phone to him.

'Pete, where are you, mate?'

'Sorry, boss,' Pete responded, and I could tell that he wasn't in a good place. 'I felt a bit under the weather today so I thought I'd stay in bed.'

'Look, Pete, if you don't come today and pack your gear, you won't be going. Come on, mate, sort yourself out. This is an opportunity of a lifetime. If you don't go now, I can guarantee that you will regret it for the rest of your life.'

Two hours later Pete arrived in a blind panic, wondering if his place on the expedition was secure. I explained that he had just suffered from a bit of cold feet, something that could happen to anyone. I assured Pete that he was an essential part of the team and he relaxed, and by the end of the day he seemed genuinely pleased that he hadn't pulled out.

We flew out to Greenland via Copenhagen, and on 4 May arrived in Kangerlussuaq, which the Greenlanders refer to as a town but in reality is a small community consisting of a couple of basic hostels, a few shops and an airstrip. The 56,000 inhabitants of the largest island on the planet live mainly on rocky outcrops on its extreme periphery. The rest of Greenland is covered by a permanent ice sheet. There are no roads linking towns, and travelling between communities is by boat, plane or helicopter.

The locals moved at their own pace and viewed us with friendly interest. We quickly learnt, though, that they were very concerned about climate change. The summers were getting warmer and the glaciers were retreating, they told us, and it was having a huge impact on their traditional way of living.

The next three days were spent in the town at the very basic Old Camp Hotel, where we prepared our kit and packed pulks, put together food and grazing bags and gathered the last of the provisions we needed for the crossing. This also meant hiring a rifle from the hotel and buying ammunition from the supermarket. The threat posed by polar bears was small, and we were unlikely to encounter any, but just in case we did, we had to be prepared.

The weather was good and on 7 May we were driven via a gravel road up to the expedition start point known as Point 660, located as it is 660 metres up on the edge of the ice sheet. There was a real feeling of nervous excitement amongst the team. Kangerlussuaq, although interesting, had lost its novelty factor pretty quickly, and we were all itching to get on the ice.

It was a relatively warm day, the sky was clear and the wind was light. Everyone applied lip balm and sunscreen – a must in the Arctic, where you can be battered by rays direct from the sun as well as reflected from the snow. The edges of ice sheets are notoriously difficult to travel across because they are areas of almost constant ice movement. The main challenge, however, especially for the first few days, was navigating our way around huge undulating ice hummocks, which slowed our progress to a crawl. It was a torturous start. Every few steps the pulk would catch on ice and wrench you backwards, and at times simply standing upright was difficult.

Those first couple of days were the most difficult and frustrating of the entire expedition. Every step took us higher onto the ice sheet, but the constantly changing terrain meant we were often shifting between skis and crampons and unable to see more than

50 yards ahead at times. To make the going just that bit more difficult, the ice was wet and soft.

I was keen to get the team settled into a routine as quickly as possible, and so I was insistent that we were all up and ready to be moving by 9 a.m. Breaks would be for five minutes every hour and the days would end at 6 p.m. I wanted to run as relaxed a ship as possible, especially as I was now leading civilians, but I was insistent that we stuck to those timelines. I led the way most mornings, breaking trail and setting a pace that was manageable, with the others following behind in single file. I thought I was being fairly laid-back, but subsequently found out that Pete and Wendy had privately dubbed me 'Loucifer'!

As well as the strict daily routine, I had another little tradition that I shared with the team, which probably drove them equally mad. Years ago, either someone told me or I'd read somewhere (and I've been unable to find it since) that it had been calculated that if Scott and his men had taken just eleven more steps each time they stopped on that fateful expedition, that accumulated distance would have meant they might well have made it to One Ton Depot, the large resupply of food and fuel they were so desperately trying to reach. Sadly, they froze to death in their tent a few miles short. This really stuck in my mind as a graphic illustration of the sometimes tiny margins between success and failure – or, in their case, life and death. So, whenever I'm on expedition now, no matter how bad the weather or how exhausted I am, once I reach the end of the planned day, I'll pause for a few seconds, and then I'll take eleven more steps, just in case it's enough to make a difference. I'm quite sure it drives my teammates nuts when they think we are done and suddenly I shuffle on a little further. Whether it's true or not is almost irrelevant now, it has become an ingrained superstition in me, and indeed I extend that whole principle of doing just that little bit more to my daily life wherever possible, whether it be preparing for a sporting event or revising for an exam. You just never know.

The days were long, with lots of halts, as I left the team to scout forward and find a navigable route. Virtually every expedition that traverses Greenland does so with an experienced guide. But in an attempt to keep the costs down, we were self-guiding, which made it feel more of a genuine adventure. Success or failure was all down to us.

Eventually – and to everyone's relief – the ice began to flatten out and progress improved. Higher up on the plateau, it was a scene of breathtaking beauty. Spread out in front of us was a vast expanse of pure, untouched virgin snow, which only stopped where the land met the sky. Everyone stopped and silently stared in awe. We were now completely immersed in Greenland and there was no going back. We would fail or succeed as a team.

After a few days, Arabella really began struggling with blisters. Her boots did not fit properly and with every step they were rubbing into her toes and heels. The pain was horrendous and she was often reduced to tears, but she refused to give up. Her spirit and guts were truly admirable. Since Chris, her tent-mate, was a trained paramedic, every evening he would spend up to an hour dressing her feet and modifying her boot in an effort to relieve the pain. But Arabella eventually reached a point where she felt she could not continue, not least because she didn't want to hold everyone back. I was determined to keep the team together, and persuaded Arabella to keep going for a few more days and reassess when we reached our first major landmark, an abandoned US early-warning station called DYE-2 – a frozen relic of the Cold War.

The warning stations stretched in a daisy chain across the top of the globe, and their primary role was to listen for the arrival of Russian intercontinental ballistic missiles. The base was abandoned in the 1980s when advances in satellite technology made them redundant. Remarkably, though, they have been preserved by the frozen, sterile Arctic air, and look almost the same today as they did thirty years ago when they were filled with US servicemen and -women and technicians, awaiting the first signs of a nuclear apocalypse to emerge out of the dark skies of Eastern Europe. When the time had come to leave the station, only the sensitive technology was taken – everything else was left behind, including a well-stocked bar and shelves full of machinery and equipment.

As we made our way towards DYE-2, the weather began to deteriorate, and on Day 6 our Norwegian fixer, Lars, warned us that a severe storm was heading our way. I'd hired Lars to help with arranging the various radio, rifle, expedition and you-name-it permits required by the Greenland authorities. He is the godfather of Greenland crossings, having guided multiple expeditions. Now in his semi-retirement in Oslo, he helps expedition groups navigate the seemingly insurmountable bureaucratic obstacles in their way. Part of his service also includes providing weather updates for the crossing.

Now Lars recommended that we didn't travel. He advised us to dig the tents in, build snow walls and prepare for a period of sustained high winds, known locally as a 'piteraq', meaning 'that which attacks you'.

I took everything he said at face value, despite knowing that weather reports are never 100 per cent accurate. When the storm eventually arrived, it didn't live up to the billing given by Lars, and as far as I was concerned we could have skied through it. The top wind speed was around 50 mph, which is strong but would not have been too much of an issue. When he asked how we had got on, I replied with a tongue-in-cheek, 'It was a bit breezy, mate.'

Three days later, Lars emailed that another storm was coming our way. To be honest I thought that he was being overcautious, and so I emailed back saying that we might try and ski through it as we were now within a few miles of DYE-2. His response was almost immediate – a strongly worded warning that the approaching storm was very serious, with wind speeds predicted at hurricane strength of up to 120 mph. He went to great lengths to tell me not to be blasé and instead make sure the camp was well prepared for the very powerful and dangerous weather front that was heading directly towards us.

It was now Day 9 and we were one day away from DYE-2, so the question was what to do. I sat down with the team and explained the options.

'A piteraq is coming straight for us,' I told the group. 'We have ten to twelve hours to get to DYE-2, which would mean that we could get inside the building and would be protected from the storm, but we would have to really push it. DYE-2 is around 16 miles away, so that is going to require a huge effort.'

I watched the faces of the team as I explained the two options. Up until now there had been an almost adventure-holiday feel to the expedition, but this was deadly serious.

'That is further than we have gone in any single day so far. But I believe we could achieve it. The gamble,' I continued, 'is that if we don't get there in time, we won't have long enough to properly prepare the tents for the storm. That would mean we could be very exposed. We could lose the tents and that would be a major emergency.

'The other option is to sit it out here. Prepare the tents as best as we can, build snow walls and make sure that we are in the best shape possible for whatever the weather throws at us. Either way there are risks. And we have to be mindful that the storm may arrive early – the ten-hour arrival forecast is only a prediction.'

The team was split. Chris and I were up for working our socks off to get to DYE-2, but the other four were clearly apprehensive and wanted to use the ten to twelve hours we had to prepare. They had a point. If we didn't make it to DYE-2 in time then we were in big trouble, and it could potentially be the end of the expedition, or worse.

The other factor that needed to be considered was that we

were not 100 per cent sure we would be able to get into DYE-2. The whole complex was and is gradually being consumed by ice and snow, and it was quite possible that the entrances were blocked. So, there was the real possibility that we could race to get there, only to find out that we wouldn't be able to take shelter inside.

It was probably the first real moment of tension in the group. Chris really wanted to go for it and was annoyed with the others, but I decided that the responsible and safe thing to do was to bunker down, prepare as best as we could and ride the storm out.

We began working in our pairs, Wendy and I, Chris and Arabella, and Pete and James. Wendy was relieved we were staying although apprehensive about the next few hours. She had an incredible character; she was the smallest in the group but in many ways had the biggest heart, and up to that point had really relished the challenge. I know she found the going difficult on some days, and had various coping mechanisms to help herself get through the difficult moments, but she never complained.

The tents were carefully positioned so that the weakest part, the door area, was downwind from the direction of the storm. All three teams excavated an area of at least two foot down, so the tent sat in a sort of pit, and each had a six-foot-high, semicircular snow wall around the top of the tent, which was designed to deflect as much of the wind as possible. It became a race against time, with a lot of snow and ice for us to shift. As the hours ticked by, I would routinely check up on the other teams to make sure they were making good progress. Unfortunately, despite my instructions, Pete and James had positioned their tent the wrong way around. It's the sort of thing that happens when people are tired and stressed, but it was no major drama.

'Guys, that tent needs to be turned around,' I said. James and Pete, who was sweating and working just in a T-shirt, both stopped digging and stared at me in frustration.

'Remember what I said – you always have the door downwind, so if you have to open it the tent doesn't fill with spindrift or the wind rip it apart. It will take you twenty minutes to sort it out but it will be worth it. Trust me.'

'Yep, OK. We'll get it done. Sorry, Lou – basic mistake.' So I left them to it.

The first signs of the coming storm began to emerge after around eight hours. The wind speed picked up and then dropped, as if it was giving us a foretaste of what was to come. Then, almost without any warning, a full gale began to blow, which just got stronger and stronger.

By around 7 p.m. that evening, everyone was bunkered down in the tents as the storm outside began to rage. The tent walls bent and buckled as the wind grew stronger and stronger. By 11 p.m. it sounded as if we were stuck inside a jet engine – the noise was incredible and pretty daunting. As the power of the storm increased, it felt as if some giant, angry wild animal was outside, roaring and banging on the tent, and for the first time in all of my polar travels I wondered whether the tent would hold. Every two or three hours I poked my head out to check the guy ropes and dig some of the snow away from the tent. The wind made it difficult to breathe and my goggles and mouth quickly filled with snow. It was impossible to stand, and dangerous to move more than a few feet away from the tent. Even though the bright red tents were only 30 feet apart, it was impossible to see the others. The spindrift being whipped up was the equivalent of a blizzard. There was nothing to do except hope for the best and try to get some sleep. Amazingly I nodded off until around 5.45 a.m., at which point I was abruptly woken by Chris, who was shaking me and saying that we had an emergency.

My head felt thick and I was momentarily disorientated. I looked up and I could see Chris kneeling above me looking deeply concerned. He was dressed in his full cold-weather gear and had snow in his thick hair.

'What's going on, mate?'

'James and Pete's tent has collapsed. Pete is missing. He could be out there somewhere. James came into our tent around forty-five minutes ago and assumed that at the same time Pete had made his way into your tent. It seems in the confusion and panic that Pete has either got lost or remained in the tent – either way, we've got a problem.'

As I began getting dressed in my cold-weather gear, I asked Chris to run through the sequence of events again. I was determined not to rush. If I was going out into the blizzard, I needed to prepare properly and think clearly. Chris explained that the guys had been asleep when the tent collapsed as the poles gave way under the intense pressure. Pete was terrified but was reassured by James that they would be OK if they could prop up the tent's roof and keep it away from their faces, which they managed to do for a couple of hours before it became clear that they were effectively being buried alive. The snow that settled on top of the tent was being compressed by the force of the wind and was setting like concrete.

Around 5 a.m., just as dawn was breaking, they both realized that they were running out of time. The problem by that stage was that the weight of snow that had formed on top of their tent made movement almost impossible. They couldn't find their cold-weather gear or get to their boots. Slowly the amount of free space in the tent had shrunk to a bubble just above their heads, but when that disappeared it was clear they were going to suffocate.

In a desperate bid to get help, it seemed that James left the tent just wearing his thermal underwear and socks. He quickly became disorientated by both the lack of light and the swirling snow that was being forced into his eyes. The wind was so strong that he was being constantly knocked off his feet, which also added to the confusion. Although he was only 30 feet away from my tent, he missed it completely and was stumbling around, quickly becoming hypothermic, before being forced to crawl around on all fours. Then, by some miracle, he stumbled across Chris and Arabella's tent.

'Arabella and I heard the zip of our tent opening and James fell in,' Chris explained. 'He was in a right state. He could barely speak, was very cold, and just kept saying his tent had collapsed. The first thing we asked him was where Pete was, and he said "with Lou". He was very confused. So we got him inside a sleeping bag and calmed him down. As he started to feel better it became apparent that James had only assumed that Pete had made it to your tent – and clearly he hasn't. Look, Lou, if you are going out there, you'll need to take a compass bearing and gear up properly. I only just made it here and we're only talking a few steps.'

By the time Chris had finished explaining what had happened, I was dressed and ready to begin searching for Pete. By now my mind was racing. I was the expedition leader and responsible for the well-being of everyone. They had entrusted me with their lives and, in all likelihood, I thought, one of them had died. I breathed slowly. I didn't allow any of the intense worry I was feeling to show on my face and forced myself to remain calm.

'Don't worry, I'll find him – everything is going to be OK. Chris, stay here with Wendy. I'm going to head to his tent first. If he's there I'll bring him back, if not I'll get roped up and do a perimeter search. But let's take one thing at a time. I'm pretty sure I can get to his tent without a rope, but if I'm not back within thirty minutes, come out and give me a hand.'

With the plan agreed, I left my tent, and was immediately hit by the ferocity of the storm. I couldn't walk upright, neither could I face into the wind. The wind often sounds fiercer inside a tent than it actually is outside, but for once this wasn't the case. It was far worse outside, the sheer violence daunting. Any form of communication with Pete short of shouting in his ear was going to be impossible. I dropped to my knees and began to crawl towards Pete's tent – it was the only way I could make any progress.

After a yard or so of crawling, I looked back and could no longer see my tent – it had disappeared in the swirling snow. A few minutes later I got to the area where Pete's tent should have been, but there was nothing there. The tent, the two pulks and the skis had all disappeared. Then, for a brief moment, I saw a ski tip sticking out of the snow. I was relieved that I was in the right location, but my heart sank at the confirmation that the tent had been buried. I checked around the ski to see if there were any other signs of their gear, but there was nothing. I was just about to return to my tent and get roped up for the perimeter search when I saw a small piece of red material, about six inches square, fluttering in the snow. It was about six feet away from the ski I was hanging onto. I crawled over to the material, grabbed it with my hand and pulled, quickly realizing I had the tail end of the tent. I dug away for ten minutes or so, discovering that the entire tent was buried beneath around four feet of hard, compacted snow. Dawn was by now approaching, but the weak light of early morning was being blocked out by the thick spindrift. Snow was being forced inside my goggles and mouth and I was struggling to see and almost breathe.

I carried on digging with my hands, praying Pete was still inside – and had survived. If he had left the tent at the same time James had, there was no chance he would still be alive. I dug away manically until I eventually found a zip. My spirits lifted as I felt some movement, and when I opened the zip I saw – at the end of a small, rabbit-hole-sized tunnel – Pete's face staring up at me. The relief was overwhelming. He was spitting snow out of his mouth and his eyes were wide and disbelieving.

'Boss! Boss! It's me, Pete! I'm here, I'm alive,' Pete shouted, but I could barely hear him.

'I've got you, mate. Stay calm. I'm going to dig you out,' I shouted at the top of my voice. But I wasn't sure he heard a word.

Another ten or fifteen minutes longer and I reckon Pete would have suffocated, much like an avalanche victim. But he wasn't safe yet. He was still inside his sleeping bag, trapped by the sheer weight of the snow, which had settled on the tent and effectively pinned him down.

Over the next twenty minutes I dug like crazy, pulling away huge slabs of compacted snow. I eventually made a sort of tunnel and was able to pull Pete out from his sleeping bag. He was seriously cold and was dressed only in his thermals, unable to reach the rest of his gear. The challenge now was to get back to my tent, which had disappeared in the blizzard.

'Hang on to me Pete,' I shouted into his ear. 'I'm going to grab your sleeping bag and then we'll crawl back to my tent.'

I bundled Pete's sleeping system under my arm – there were no spares, so it was vital his bag came with us – and we set off, crawling like animals in the snow into the ferocious storm. We had only gone a few feet when Pete was suddenly ripped away from me by a massive gust, and barrel-rolled off sideways into the blizzard. I looked around and could just make out his silhouette, but Pete was now disorientated and began crawling off in the wrong direction.

'Pete, Pete, this way,' I shouted as loudly as I could, but the roar of the winds silenced my calls so I quickly crawled after him, stumbling and falling but just managing to catch the back of his legs before he completely disappeared.

Pete was now in a real state, but there was no time for niceties.

'Pete,' I shouted about an inch away from his face. 'Don't let go of me. Whatever you do, don't let go of my arm. We've got to get back to my tent.' He nodded and I was relieved that he understood, but I could also see that he was pretty terrified and confused.

It seemed to take an age, but I eventually managed to get Pete back into my tent. Chris and Wendy were hugely relieved, but

Pete was very cold and pretty shaken up. We quickly got him into a sleeping bag and I began to mentally assess what had just unfolded. I couldn't understand why Pete's tent had been the only one to collapse, and I began to consider the options for the rest of the expedition.

As Pete began to calm down, he sheepishly admitted that he and James had decided not to follow my instructions to position the tent door downwind. He apologized profusely, but now was not the time for regrets. He also explained that they had hoped to remain in the tent until the storm subsided, but when it became clear that they were slowly being trapped under the heavy weight of snow, James left the tent and Pete thought he had gone for help.

Pete, who was now warming up, continued: 'After about half an hour, when no one had turned up, I had a look outside and knew immediately that if I left my tent I would die. So I thought I'd stay in the tent – I knew that you would eventually come and get me, boss. Then I began to wonder whether James had managed to actually get to another tent. It reached a point where I was trapped. Because of the amount of snow on top of the tent, I could barely move, and that's when I really began to worry. I thought I would suffocate. So, I did what any man would do and reached for my bottle of Jack Daniel's. I thought, if I'm going to die, then I'm going to go out in style. I also recorded some thoughts for my family on my iPhone. My final message to my kids, so to speak.'

Pete really had thought that his time had come, and had been preparing for the end. I wondered whether he was going to have the mental strength to go on. In fact, I wondered – after the trauma of the storm – whether anyone would want to continue.

Chris went back to his own tent, pinged me a message when he got back, and then we all settled down for the next few hours.

Several hours later, the wind had died down completely, and we left our tents to survey the carnage. Everything had been covered in hard, compacted snow, including the tents, which were partially buried. The pulks, skis and poles had disappeared completely. James and Pete's tent was wrecked, but it still needed to be recovered, along with the rest of their gear.

'OK. Everyone is aware of what happened last night, so there is no need to dwell on it,' I told the team. 'We are still on for the crossing. Pete will come into my tent and James will share with Arabella and Chris. It will be a bit cosy, but I'm sure it will be fine – they are three-man tents, after all. We now have to dig out all the equipment, which is going to take a while, and then we need to get going again and head towards DYE-2 as soon as we can.'

I could see immediately that there was a lot of tension between Pete and James, and the last thing I wanted after such a traumatic event was bad feeling among team members. So I had a quiet word with James, who was the more experienced of the pair, and explained that we needed to work together as a team. It took six hours of hard work to dig everything out and I was happy that everyone was busy – it meant that the team had something constructive to focus on rather than dwell on how close we had come to disaster.

As we set off later that day, my mind was occupied with how we might cope if we were struck by another piteraq. Last night Pete had nearly died. We had also lost one tent and team morale had been fractured. Another major storm might just bring the expedition to an end. But I kept those thoughts to myself and ploughed on the following morning as if it was just another day.

Gradually, the eerie site of DYE-2 became visible on the horizon. It must have been a strange experience being stationed at the base as a US serviceman, thousands of miles away from home and knowing that – if you ever had to do your job for real – the world was probably going to end in a storm of nuclear explosions. We arrived at the vast structure, unharnessed our pulks and walked up a steel stairway to a small gap through an open door. We entered a hallway and then a wider room, which was lit by sunlight streaming through an array of windows.

Walking inside felt surreal. It was as if the personnel based there had literally been told that they were leaving that morning. Everything had been left behind. There was beer in the fridge, books on shelves, magazines in bedrooms and even fake flowers on the tables.

As we walked through the complex, we were greeted by members of another expedition from New Zealand. They had arrived a few hours earlier, having left Kangerlussuaq two days before us. The team was the same size as ours, with two females, but was being led by the highly experienced Norwegian polar guide, Bengt Rotmo. Bengt had completed this crossing twelve times to date.

They too had survived the storm, and Bengt was amazed, given that none of our team had trekked across Greenland before, that we had managed to escape reasonably unscathed. That evening we socialized together, chatted about the storm and the challenge of the next couple of weeks, especially as there were reports coming in of deep, soft snow on the eastern side of the ice sheet.

Bengt's team left earlier than us the following morning; over the next few days we would catch up and overtake them, then they would do the same with us. We weren't racing each other, it was just the way events panned out. As we got closer to Greenland's east coast and our finish point, the going became very tough. There had been huge snowfalls and in some places we were ploughing through soft snow up to three feet thick. Every step was a real effort, with the soft, wet snow sticking in huge clumps to the skins on the bottom of our skis. This was caused by a temperature differential between the air and the air trapped in the snow. Additionally, the pulks were sinking beyond their runners, creating extra friction.

Several expeditions who were ahead of us had pulled out, but our morale was high and we ploughed on. Meanwhile, the New

Zealand team were now about a day behind us and were struggling. They had reached the point where Bengt was seriously
considering packing it in when, purely by luck, they came across
our tracks. They dropped into our cut track and eventually
caught us up a couple of days later.

There was about four days of hard skiing left to do, and Bengt
proposed that we work together and share the load. It made perfect sense to me and the rest of the team. We all knew that it was
going to be a very tough last few days, and anything to lighten
the load suited us just fine. We were now a twelve-person team,
which meant there were more people able to go up front, breaking trail through the soft surface. Bengt also explained that the
last part of the route was quite technical and he was happy to
lead – which again suited us all. The last four days were great
fun, and it was good to have some more people around, especially in the evenings when we had a short time to relax. We
finished off with a marathon twenty-hour ski session to make the
helicopter pick-up point high above the town of Isortoq. Finally
stepping off the ice sheet onto solid rock was a monumental
moment for the team.

As with all expeditions, the ending came quickly and abruptly.
The months of training, planning and preparation, and then the
expedition itself, all seemed to have flown by, and before we
knew it we were flying home.

Even before we landed in Iceland, my mind began turning
towards the South Pole expedition – which was looming ever
nearer – and my complete failure to make any meaningful headway with the fundraising.

Shortly after checking my bags in for the flight from Iceland
back to London, my phone rang. It was my boss, the commanding officer at the Infantry Battle School in Brecon. I immediately
felt a surge of panic. Because of some flight delays at the end of
the trip, I was slightly overdue returning to work. But I needn't
have worried.

'Lou, it's Shaun Chandler.'

'Hello, boss,' I responded. 'What's up?'

'I've got some good news. You've been awarded an MBE for your leadership on the SPEAR expedition. You should be very proud. It's an amazing achievement.'

I'm not often lost for words, but I was at that moment. SPEAR had been a great project to have been part of and that had been enough of a reward for me – you never enter into these things thinking of any recognition. So to receive an MBE was totally unexpected and a huge honour. I would treasure it as a permanent memento of an amazing couple of years. Lucy and the children were absolutely thrilled too when I called them with the good news.

14

A MINOR MIRACLE

This is no time for ease and comfort.
It is the time to dare and endure.

WINSTON CHURCHILL

I had only been home from Greenland for a matter of hours before I restarted the planning for my solo Antarctic expedition. It was the beginning of June 2018, and the expedition was due to start in five months' time. The workload was huge, and I felt a mild sense of panic rising within me. Almost immediately I began to realize the full impact of what a solo expedition was going to cost me personally. I was used to working in a team, where the burden is shared and everyone works to a common goal. That was life in the Army and SAS, and it had been the case with the two previous expeditions where I had delegated various bits of the planning to other team members. That strategy had allowed me to manage, finesse and fine-tune the expedition while making sure everything came together. But now everything was on my shoulders and the pressure was immediate.

Before leaving for Greenland, I had submitted the initial expedition application form via Army Adventure Training. It was essentially a business plan for an expedition, where I had to provide details of funding, together with a risk assessment and my thoughts on how the expedition would benefit the Army. I also had to provide a media plan and explain how medical issues would be covered – like who would come and rescue me if everything went wrong. That application form was fortunately already bouncing around the Army, which gave me a bit of breathing space, but not much. But the main challenge was fundraising. Somehow, I had to find £150,000 in the next few months, otherwise it was game over. I needed to find a unique angle to attract sponsors, who had to be convinced that I would succeed – no one wants to be associated with failure! But Lady Luck was on my side.

Firstly, Ian Holdcroft and Martin Brooks, the co-founders of Shackleton, a relatively new company producing expedition-grade clothing, wanted to come on board, not just as sponsors but as expedition partners. The support package they offered was incredible and went way beyond just funding, including media and marketing and the opportunity to use their bespoke specialist clothing, which performed brilliantly during everything Antarctica threw at it. The whole package was worth in excess of £50,000. For me it was the perfect partnership with a company whose values and vision were deeply rooted in the ethos of a polar legend.

Secondly my boss, Lieutenant Colonel Shaun Chandler, completely bought into my plans for the solo expedition, and told me to take as much time as I needed to train, prepare and plan. He is a top bloke and I couldn't have wished for a better response.

Colonel Chandler was the first level of approval, but my plan now had to go further up the chain of command. The Army classifies a polar expedition as Level 3, High Risk and Remote (HRR), which means they viewed it, and rightly so, as being seriously risky. There's no Level 4! The next challenge was convincing Major General Paul Nanson – the general officer commanding Army Recruitment and Initial Training Command – to be what the Army calls the 'ODH', the overall duty holder. That meant if it all went wrong, he would be asked to explain why.

As far as the Army were concerned, the expedition planning

process had to pass the *Daily Mail* test, as they called it. If I was injured or died, was the Army going to face criticism? The backdrop to all of the extra concern was Henry Worsley's death. Henry had no longer been serving when he died, but his death was still fresh in everyone's mind and had become a sort of unspoken reference point. The tacit view was that if Henry, a man of huge strength and polar experience, could succumb to the conditions, then so could I. If something went wrong the front pages could read 'Second Army Officer dies on South Pole Expedition' – with various criticisms being levelled about a lack of duty of care. Nobody openly said that to me, but I sensed that was a major concern.

But the meeting went well, and Major General Nanson came across as hugely supportive. In August I received confirmation that the Army was prepared to underwrite the total cost for the expedition. The understanding was that I would still try and raise as much as I could through my own efforts, but the Army would cover any shortfall there might be. That was a huge relief and eased the pressure enormously. It meant that I could start buying equipment and food and fine-tuning the planning, knowing with a degree of certainty that I was actually going.

While I was preparing for meetings and writing up presentations, I was firing off sponsorship letters to various organizations in the hope that they would offer up some financial support. I was also travelling the country giving presentations to companies I had targeted as potential sponsors, only to find out a few weeks later that they had no money left in the budget. It was a hugely dispiriting process, and there were times when I seriously considered postponing the expedition for another year. It was around this time, in late May 2018, when I started struggling with sleep. I had always been one of those people who could happily sleep on a rock, but now I would suddenly wake up at 3 a.m., almost in a panic, worrying about the expedition failing and all the different things I still had to do. I often struggled to get back to sleep and felt the need to start jotting down thoughts on my iPad, just in case I'd forgotten them by morning.

Slowly my letter writing and grant applications began to bear fruit, and the funds started to trickle in. It became a matter of personal pride that I was able to mostly fund the trip through my own efforts and, after a tricky conversation with my wife (the 'long-haired general' as she was affectionately known), was able to make a substantial personal contribution from our limited savings.

Then, just when I began to relax, questions began to emerge about the medical safety plan. The problem was that the conditions I was going to face in Antarctica were in excess of the guidelines for military cold-weather training. Below a certain temperature, all training should cease, and if I sustained any kind of cold injury, Army policy dictated that I should be removed from the environment immediately. In reality, most of the expedition would be conducted beyond the temperature guideline, and a bit of frost nip is not necessarily a show-stopper, as I had discovered on the trip with Henry.

The other issue for the Army was that ALE couldn't provide an actual guarantee that they could rescue me quickly if I was in trouble. In Iraq and Afghanistan, the military worked on the basis that a severely injured casualty needs to be off the battle-field and into an operating theatre within an hour for the best chance of survival, a period known as the Golden Hour. There were no such guarantees in Antarctica. The unpredictable nature of the weather, even in the summer season, meant that a rescue could take days. It was part of the deal.

The days were now ticking by and I could feel the anxiety building inside me. I realized the medics were only doing their jobs and they had justifiable concerns, but it seemed that at every meeting another potential issue was thrown up. They insisted that I had an abdominal CT scan just to make sure I didn't have a pre-existing condition that might manifest itself during the expedition. They insisted that I had one-to-one training with a doctor who had experience of extreme cold-weather injuries, and that I should also have a dive medical – the most comprehensive medical that can be undertaken in the Army. It was essentially a full MOT, looking at my heart-lung capacity and blood pressure. But I put my foot down when someone made the suggestion to General Nanson that I should consider having my appendix removed just in case it burst during the expedition. It had been fine for the last forty-nine years, so I elected to keep it, although I suppose it would have saved a bit of weight.

The net effect was that by September I was seriously frazzled, to the point where I put an email out to everyone involved stating that – unless the problem-finding process ended – I was going to arrive at the start of the expedition mentally exhausted. The expedition had become all-consuming and the stress was beginning to take its toll on me.

While all the medical issues were being resolved, I turned my attention to the equipment I would need. I wanted to have as much new gear as possible to reduce the chances of equipment failure. I acquired a new, lighter, smaller tent from Hilleberg, a longer, wider pulk from Acapulka, a more efficient down sleeping bag, new boots and skis. My clothing was largely new, too, with most of it coming from Shackleton. I decided not to have any luxury items, such as a book; instead I downloaded thirty-five audiobooks on to my iPhone, as well as around 5,000 songs. A lot of the audiobooks were about military history, including the superb *SAS Rogue Heroes* by Ben Macintyre, and quite a few biographies of Winston Churchill. I also had a few movies and back-up data covering the waypoints for the route I was taking and details on the contents of the medical pack, as I had binned all the medication packaging to save weight.

As far as communications were concerned, I would be using

an Iridium satphone,* and my back-up was a small device called an Iridium Go, which gave a data link to the satellite and which I was going to use to send emails, very low-resolution photos and limited voice calls. Everything electronic was rechargeable using a lightweight solar panel; I was looking to maximize the twenty-four hours of daylight that the summer season in Antarctica offers up.

I produced a spreadsheet of all the gear I needed, and the individual weight of every single item. The total weight was going to be around 160 kilograms, including the pulk. To my dismay I knew that it was too heavy to make the kind of daily mileage I needed. I've never had particularly strong legs, even earning the nickname 'chicken legs' while in the Marines, and I've always had to work hard in the gym to compensate for this. Therefore, getting the weight down became a key priority for me.

So I set about doing whatever was necessary to reduce the weight. I sourced lighter carbon-fibre options for my snow shovel and cooker board, I cut away bits of extra metal from the cooker, took all the labels and unnecessary bits from my clothes, and removed metal zipper pulls off the tent. I also planned to empty all of my dehydrated meals from their foil packets into lightweight freezer bags when I arrived in Chile. That alone was going to save me around six kilos. It was an ongoing – and seemingly endless – process. I looked again and again at every item and asked myself what I could do without. Did I really need two sets of gloves, hats and face masks, or could I make do with one set? I also elected to do the whole journey in only one set of underwear and thermal base layer. In which case I needed to make sure there were no toilet accidents, which frequently happens on polar journeys while fumbling through multiple layers of clothing in thick gloves.

---

* At the time of writing, Iridium are the only network providing full coverage of Antarctica.

I also decided that I needed to do a bit of calculated risk-taking. My main skis would be a pair of Åsnes expedition-grade skis, but my spares with bindings on would be lightweight mountain race skis, which would save a few kilograms in weight. The only other items I doubled up on were ski poles and a spare cooker. Both were show-stoppers if they broke. Losing the tent in a storm would also have spelt the end, so I carried spare poles and an emergency bivi shelter for this eventuality. For everything else I had an extensive repair kit with glues, tape and plasti-ties.

I eventually reached a point where I had managed to reduce the weight down to 140 kilograms in total. It was a huge saving and helped psychologically to know that every gram I was hauling behind me was critical to the success of the journey; there was nothing superfluous.

At this point ALE's legendary UK rep Steve Jones told me that most of the other people applying to traverse Antarctica at the same time as me had dropped out. For reasons of confidentiality, he couldn't give me the name of the only person remaining. I made a few enquiries, but no one within the polar world seemed to know who he or she was. I wasn't too bothered because, frankly, I had enough on my plate getting my own expedition sorted.

Between returning from Greenland in June and departing for Chile at the end of October, I did not take a day off from planning. To be honest, I doubt that I took an hour off. I would sometimes wake up at night with an idea about how I could improve the equipment, or have a brainwave about fundraising, and the rest of the night's sleep would be lost as I began making notes.

As well as all the planning problems, I sometimes had to deal with the doubt that I actually had the ability to ski across Antarctica. Henry was probably one of the toughest individuals I had ever come across, and yet the journey had cost him his life. I had seen him in action in Antarctica first-hand, and experienced just how comfortable and capable he was in that environment. By comparison I regarded my own performance on that journey as somewhat lacking. On the SPEAR expedition, when I had been the leader, I had had to encourage, cajole and lead a team. Their lives had been in my hands, and I had felt entirely responsible for ensuring that they made it safely through. But a solo expedition is a different story. Was I going to be able to drag myself out of that sleeping bag every morning and ski in the most appalling conditions for twelve hours, and then do it again the next day, and the next?

It's important to note that there had been previous successful solo crossings of Antarctica, the first and most notable by the legendary Norwegian explorer Børge Ousland in 1996–7. He started close to Berkner Island and crossed most of the Ronne Ice Shelf, passed through the South Pole and finished on the outer edge of the Ross Ice Shelf at McMurdo. He did the crossing with no food resupplies, but used a kite to harness the wind, and so was able to cover 1,864 miles in just sixty-four days, averaging over 29 miles a day. A truly inspiring and groundbreaking expedition. He is a genuine hard-core explorer with some incredible accolades and I admire his achievements immensely.

Likewise, Mike Horn in 2017 was able to travel over 3,100 miles in just fifty-seven days, averaging over 54 miles a day. In 2019 the forty-nine-year-old Dr Geoff Wilson was able to beat this, covering just under 3,300 miles in fifty-eight days, averaging nearly 57 miles a day. He reached speeds of over 30 mph and on his best day covered 130 miles. The best I had managed in a single day of manhauling had been 19 miles, so the benefits of kiting were obvious. A rough comparison with kiting and manhauling could be the difference between sailing across the Atlantic Ocean and rowing across. They are very different types of journey.

The few other successful solo attempts had all either used

kites or food resupplies. I was planning to go 920 miles unsupported and rely on muscle power alone, but crossing just the Antarctic landmass and not including the full ice shelves. I'd be starting and finishing on their inner edges. It was a sensible goal for my first foray into solo travel.

I'd never done anything solo in my professional life, and my first solo undertaking was to attempt something that had never been achieved in the most hostile place on earth. The thought literally sent shivers down my spine. Naysayers within the polar community began commenting on various forums that the crossing couldn't be achieved without a resupply or use of a kite. And there was some truth in what they were saying. Stronger, fitter and more experienced people than me had failed. But I had to believe in myself. I was stronger, fitter and more experienced than I had been on previous expeditions, and I had been in the SAS for twenty-five years – that, for me, was the difference. Mental resilience was going to be the key to cracking this journey.

Then, out of the blue, on 18 October, I received an email.

'Hi, Lou. My name is Colin O'Brady and as you may or may not have heard I am attempting a similar project to you in Antarctica this season. I just publicly announced my project today and wanted to reach out to you directly as our journeys will certainly be linked in some capacity.' He then went on to suggest that we arrange a call to talk about logistics.

I sent a short reply suggesting that it might be better to meet up in Chile given that I was under significant time pressure.

I hadn't heard of Colin before, but a quick search online revealed that he was a thirty-three-year-old American professional endurance athlete who had once represented his country in the triathlon. He was a serious contender. He had undertaken some major endurance events and notched up a string of world records in mountaineering.

If I'm being completely honest, I was briefly annoyed by

Colin's email. I suppose I felt somewhat as Captain Scott must have done a century earlier, when Amundsen changed his plans last minute to focus on the South rather than the North Pole, and sent Scott a message that simply read: 'Beg leave to inform you, *Fram* proceeding Antarctic, Amundsen.' History was repeating itself and I sensed the media were going to have a field day.

It struck me, though, that our two expeditions were very different. Colin's was called 'The Impossible First' and it was clear that he was intent on becoming the first person to traverse Antarctica solo and unsupported without the use of a kite. He was a professional adventurer and this was his full-time career. I thought it was a risky proposition calling his expedition this, as he immediately placed himself under immense pressure to be first, and was it really impossible? There had only been two previous attempts (Ben Saunders and Henry Worsley) at this type of solo, unsupported, landmass-only crossing. It seemed to me that it was just a matter of time before someone cracked it.

Sure, I too wanted to be the first person to traverse the South Pole, but my expedition – 'Spirit of Endurance' – with its obvious historical links to Shackleton's attempt to traverse the continent, came from my love of Antarctica and deep interest and respect for polar history. I was no professional adventurer; I was a soldier first, and very much, in my opinion, an amateur in the world of polar travel, much as Scott and Shackleton had been. I did this as a hobby, for the love of the journey and the personal challenge, not for any perceived recognition or world title.

The last few days before I left were something of a groundrush. Shackleton organized a big, swanky expedition launch event at BAFTA in London, to which all the sponsors, friends and family were invited and where they were thanked for their help and support. On the morning of the launch I also picked up my MBE for the SPEAR expedition at Buckingham Palace, which was presented to me by Princess Anne, who laughed when I explained that I was about to depart for Antarctica once again. It was a wonderful day and I was able to share it with Lucy and our daughters Amy and Sophie, although sadly Luke was away with the Marines at the time.

By that stage I was just itching to get on the plane and head south. My head was still spinning with all the planning, preparation and demands of the sponsors, and I was actually craving the relative simplicity of life on the ice: ski, eat, sleep, repeat.

15

THE GREAT WHITE QUEEN

'Fortitudine Vincimus' – By Endurance We Conquer.

SHACKLETON FAMILY MOTTO

On 25 October I slumped into my economy-class seat on the first leg of my flight to Punta Arenas on the southern tip of Chile, closed my eyes and almost immediately fell asleep. The relief I felt at finally being free from the endless emails, questions, and mind-boggling but necessary Army bureaucracy was palpable. I hadn't slept properly in months, but at that precise moment I felt the weight of worry and stress almost lift from my shoulders as the aircraft climbed into the sky. There was no going back now.

Twenty-two hours later I arrived, bleary-eyed, in Punta Arenas, excited to be back once again. My body clock was all over the place. I fell into a taxi and just about managed to direct the driver to my hotel, where I crashed for the next few hours. The following morning I headed straight for the ALE warehouse and began checking to make sure all my airfreighted equipment had arrived safely. The next few days would be a vital, if not particularly glamorous, aspect of the expedition. I began the long and tedious process of emptying all my dehydrated food into little freezer bags as part of my extreme weight-reduction strategy. I devised an ingenious way of getting three meals into one bag by emptying one meal into a corner then twisting it off, then adding another meal, twisting off the opposite corner and then adding the third meal on top. It was all about saving grams in weight. On a bad day on the ice, I would want to know that there was nothing more I could have done to reduce the weight in my pulk. Fortunately, Wendy Searle had again volunteered to be the expedition manager. She had joined me in Punta Arenas and her input was extremely helpful. Together we worked from 9 a.m. to 11 p.m. carrying out the final checks, packing and repacking all my equipment and chasing down the last bits of kit I would need for the expedition, such as lighters, lithium batteries, cheese and salami. At the same time I was often travelling up to the main ALE office to get weather reports and be given various route safety briefs. I also had to go through my route planning with their Travel Safety team and conduct GPS and comms checks.

A day or so after I arrived in Chile, I received a call from a *New York Times* journalist called Adam Skolnick, who had first contacted me back in the UK some months earlier. During our early conversations I'd been intrigued by his questions, which were very technical, and by the fact that an American journalist was so interested in a British expedition. I was a little cautious about releasing information that could aid a potential competitor.

He had flown into Punta Arenas to cover the story of mine and Colin's attempts to traverse the continent. Adam was a gifted storyteller and obviously keen on the race aspect, and I guess a little frustrated that I went to great lengths to stress that I was not approaching it as a race. But he was nothing if not persistent, and pushed for me to meet Colin and for the event to be recorded and photographed. I had no real objection, and later that evening I met Colin for the first time.

As soon as I walked into the bar, accompanied by Wendy and three members of the ALE team, Colin and I made eye contact. He looked every bit the athlete, and younger than his thirty-three years, and I noticed straight away that he was drinking fruit juice.

'Colin,' I said with a friendly smile. 'I'm Louis Rudd. Great to meet you.'

We chatted about nothing much, the way you do when you are trying to make conversation, but in reality you'd rather be somewhere else. Colin seemed nervous and on edge, and it later occurred to me that maybe he had expected to be given a hard time from me. But he relaxed as the evening wore on and we got on fine. He was off the beer, whereas I was determined to have a good drink, given that it would be over two months before I would enjoy another drop of alcohol.

He spoke about his early exploits and how he was looking forward to getting on the ice. He studiously avoided using the 'race' word. But the awkward atmosphere reappeared when one of the ALE team asked Colin directly why he was refusing to commit to a specific route so late in the day. All anyone seemed to know at that stage was that Colin was going in the opposite direction to me, which meant a short, sharp ascent up through the Transantarctic Mountains followed by a long, slow descent across the continent. It was a route that was far more expensive – costing at least 100,000 USD extra – and way beyond my price range. But Colin had some pretty serious sponsors and cash was not a problem. It also became clear throughout the evening that the *New York Times* journalist was going to write that Colin and I were in competition with each other, despite my insistence that I was not interested in racing.

'So, Lou,' Adam said while scribbling notes on a pad, 'do you think a Brit is going to be the first to conquer Antarctica?'

I sighed deeply.

'Look, Adam, it's not a race,' I said with barely disguised frustration. 'Firstly, we are starting from different sides of the continent, and secondly, it will be nothing short of a miracle if we both finish a journey that has never before been completed. The last thing we should be doing is thinking about racing each other.'

The evening eventually fizzled out and I woke the following morning with a thick head and a message from ALE stating that my departure time was being brought forward by twenty-four hours to 31 October due to a clear weather window. We either went early, ALE said, or faced a long delay. The rapid change of plan was a bit of a pain because it meant that I had lost a day of preparation, but on the plus side it meant that I could crack on with the expedition sooner.

On the afternoon of 31 October 2018, I arrived in Union Glacier, Antarctica. It was one of the first flights of the season and it was mostly ALE staff on board; the only expeditioners were me, Colin and an Italian ex-special forces guy called Danilo Callegari who was there to ski solo and unsupported to the South Pole, climb Mount Vinson (the highest mountain in Antarctica) and skydive over Union Glacier.

It was good to be back, and I was welcomed like a treasured guest by many of the ALE staff. The base was a real hive of activity, with lots of preparations being carried out for the forthcoming season, and while I was happy to chat, I was also keen to put my pulk through its paces and try to get an idea of just how tough it was going to be dragging something weighing 140 kilograms across Antarctica. Almost from the very first step, I knew it was going to be extremely difficult. The pulk barely moved. It required all of my effort just to take a few steps. My intended start point was Hercules Inlet on the Ronne Ice Shelf, the closest location to UG and therefore the cheapest option. But there was another option – the Messner Start Point, named after the famous mountaineer Reinhold Messner. It was further away from UG than Hercules Inlet, but it provided a shorter crossing by some 80 miles. That meant I could effectively remove ten days' worth of food from my pulk and save myself ten kilograms in weight – it wasn't much but it was enough.

Later that evening I explained my thoughts to the ALE field operations manager, Tim. He was pretty cool with the idea but said that the change would add a hefty 11,000 USD to my bill due to the extra flying distance involved.

'Let's go for it,' I said. 'I'll find the money later,' and secretly hoped that – if needs be – the Army would pay the extra.

'The other good news, Lou, is that we can get you to your start point in the next day or so. We'll have a better idea tonight. Why don't you go and get yourself sorted out and we should have an answer by dinner.'

I walked back across the ice to my tent feeling warm inside, even though the temperature was nudging -15°C. For the first time in a long while, I felt as though I was finally getting the rub of the green. As I had hoped, once on the ice, things were coming together. Just one more day and I'd be on my own, skiing south. Inside my tent, I climbed inside my cosy down sleeping bag, closed my eyes and slowly drifted into a deep sleep.

An hour or so later I was woken by a member of the ALE team to say that Tim needed a word in the admin tent. I assumed it was another updated weather report, but instead Tim explained that Colin was also starting from the Messner Point because the weather at the Leverett Glacier was bad, and a flight there would be delayed, possibly by as much as ten days.

'So, if you're happy, we could fly you both out to Messner tomorrow. The good news for you is – because you are sharing the flight – your costs will be much lower.'

'Look, it's no problem. I'm more than happy. I really just want to get on the ice now.'

I was starting to wonder if I was about to embark on a cursed expedition, and whether the endless string of problems I was now beginning to encounter was payback for trying to cram too much too quickly into the planning stages. Time would tell.

The following morning, Colin and I loaded our gear into the small hold of the Twin Otter ski plane before we said our good-byes and climbed aboard, both of us no doubt full of nervous anticipation. The plan was for the aircraft to land at the start point, drop one of us off and then taxi out on its skis a mile parallel and drop off the other. I had offered Colin the option of getting off first, which he seized.

The Twin Otter climbed into a cloudless blue sky and, as I watched out of the window, I felt my love for the Antarctic return in waves. It really was a unique place.

The flight to the drop-off point was made mainly in silence. Both Colin and I were lost in our own thoughts and then, just before we were due to land, the pilot came up on the intercom to tell us that the drop-off had been aborted. There was a blanket of thick ground mist covering the landing point, and the pilot had no view of the landing area. There was no other choice but to return to UG. Another setback, and again I wondered whether this was all part of the unravelling process. But I kept my thoughts to myself.

It was another two days before the mist lifted and finally, at around 2.25 p.m. on 3 November 2018, I climbed down the steps of the aircraft and onto the ice. Finally, I had arrived at the Messner Start Point. The ice shelf is a vast blanket of snow and ice 1,000 feet thick that has lain undisturbed for aeons. One edge is physically attached to the Antarctic continent, while the other drops like a sheer cliff into the sea. My world was now dominated by two colours – the dazzling white of the ice shelf and the blue of the sky. Ahead of me was an undisturbed horizon of snow and ice, and it would take several hours of hard hauling before I reached the continent proper.

I watched as the aircraft became airborne, sending up a thick cloud of fresh snow. A mile to my right I could see Colin, a black, shapeless figure. For the next twenty minutes I sorted out my equipment, checked my skis, harness and pulk, ensuring I was happy with my thermal layers, finally fitting my compass around my neck before taking my first step.

Glancing over at Colin, I saw he was almost bent double, like a man burdened by a heavy weight, as he dragged his laden pulk across the ice. A part of me would have found it satisfying to speed away, leaving him in my wake, but I was determined to stick to my plan – and besides, he had his work cut out and wasn't worrying about me. And he wasn't the only one suffering. Almost immediately I began to feel beads of sweat trickle down the sides of my head and neck. I was soon breathing heavily as I pushed one ski in front of the other. Every step required the utmost effort and, as my boots rubbed on my heels, I knew that blisters were coming. The next few days were going to be very tough, possibly far harder than I had thought. I had trained hard, dragging a tyre over countless miles through long grass, but nothing can really replicate dragging a 130-kilogram pulk across soft, unbroken snow.

I stopped after an hour, threw a handful of trail mix into my mouth and scanned the horizon. Antarctica was truly stunning, I said to myself, and I knew that I would never tire of its beauty. I plugged my headphones in my ears and set off again, skiing on a gentle but noticeable incline. By 7 p.m. that evening I reached my campsite and briefly stopped. I then took eleven more steps. The wind was relatively light and so I was able to get the tent pitched within fifteen minutes and begin the process of melting ice. I was going to follow the same routine every evening as I had done on previous expeditions, but now everything would take much longer because I was on my own. After rehydrating my meal and drinks, I wrote up my diary, called into ALE and filed some notes for the blog. I then finally hit the sack around 11.30 p.m.

A good night's sleep left me fully rested and keen to get on the move again. I began the lengthy process of melting ice for my porridge and hot chocolate – I needed the calories. While the ice was melting, I began packing away my equipment in the pulk, putting the heavier items at the end in the hope that it might make pulling a tad easier – it was all about marginal gains.

Within minutes of setting off, I was sweating heavily. The weather was seriously mild for Antarctica, around -10°C and virtually no wind. Within twenty minutes or so I had stopped and stripped down to my thermal base layer, a thin merino wool top. The pulk was heavy, almost too heavy, but there was little point in worrying about it – and besides, from now on it would get lighter as I ate my way through seventy-odd days' worth of freeze-dried rations. At the end of one hour I stopped for a breather and could see Colin in the distance. His posture was not good and he appeared to be struggling with the weight – just like me – and I wondered how long he could last skiing in such an awkward position.

Those early days were a tough test for both of us. Colin was younger, fitter and stronger than me, but I was the more experienced campaigner. What was clear, though, was that he was a seriously determined character, and sometimes, in Antarctica, that can be enough.

As lunchtime approached, a stiffening, icy wind barrelled down the ice shelf and hit in what amounted to a full-frontal assault. Within minutes my body temperature had dropped to the point where I was beginning to shiver. I stopped and quickly pulled my down jacket from my pulk just in time to stop the sweat freezing on my back. I ploughed on, with the wind blasting my face, as if I was skiing in my own personal wind tunnel. Around half an hour later I was sweating again and had to remove my jacket, mitts and hat, and so my day continued – warm kit on, warm kit off. With each slide of my ski and tug of the pulk, I felt the energy drain from my body.

By 3.45 p.m. I was exhausted and decided to call it a day. I barely had the strength to pitch my tent and was too tired to begin my food prep, so I removed my boots and watched the steam rising from my sweat-soaked socks. I was almost overcome by a sense of pure ecstasy as my feet began to breathe again. Removing my boots was the best moment of every day and the one thing I would focus on when the going got really tough. Only another hour, I would tell myself, then boots off. I 'washed' away the sweat with some snow, dried them thoroughly with a small hand towel and then put on my bed socks. I just about had enough strength to climb into my sleeping bag and quickly fell into a deep sleep.

I woke two hours later, disorientated and blinking into the watery light of the tent's gloomy red interior. I remained stationary for a few seconds, enjoying the moments of peaceful solitude and deciding on what dehydrated meal to have. Food in the Antarctic is more than merely fuel. It boosts morale and provides an aiming marker for the day. After breakfast it could often be another thirteen hours before my next hot meal. I had a rolling five-day menu of chilli con carne with rice, chicken tikka masala, spaghetti bolognese, Asian noodles with chicken, and spaghetti carbonara, and every tenth day I had a chocolate pudding. The chocolate pudding was all part of my mental coping strategy. It was something to look forward to and, because I had my food packaged in ten-day sacks, I knew that every time I reached a chocolate pudding, my pulk would be around 15 kilograms lighter from the consumed food and fuel. It was like a treat, or a motivator – get through another ten days – and it worked really well. My only other food luxury was a small pot of curry powder, which I would occasionally sprinkle onto my food, just to give it a bit of extra lift. Then, after dinner, I began preparing my audio blog entries, wrote up my personal diary and sent a sitrep to ALE.

Over the next two days I clocked up 22 nautical miles and felt as though I was making good progress, but at the same time my heels were taking a real battering. I knew, even at that early stage, that my feet were going to suffer. In the past I had tried just about everything to prevent my heels blistering, but nothing had ever worked, and now I just accepted that I had blister-prone feet. I also caught the occasional glimpse of Colin battling away behind me, his body bent almost in two as it had been on day one. It seemed to me that he was following in my tracks, literally, but it didn't worry me. I thought it was a sensible move and I probably would have done the same if I was him.

On the evening of the fourth day, as I lay on my sleeping bag, airing my feet and chugging a litre of strawberry-flavoured protein drink made from ice I had just melted, I began to weigh up the advantages and disadvantages of a solo journey. Before leaving the UK, I had wondered how I would manage seventy days of solitude with nothing to keep me company apart from my music, audiobooks, and the voice inside my head. But the reality of those early days was that I was too busy to feel lonely. There was always plenty to keep me occupied in the evenings. More often than not there would be a minor kit issue that needed to be sorted, or reports and diaries to be written. It was rare for me to actually get to sleep before 11 p.m. In fact, the major downside of a solo expedition was the increased workload – everything took twice as long and that had to be factored into the evening routine. On the SPEAR expedition, and earlier with Henry, we had shared everything from pitching the tent to the cooking and even writing up the blog. But by far the toughest part of a solo journey was being on the ice on your own and having to cut a fresh trail all day, every day. It was a killer.

On SPEAR everyone would lead for seventy-five minutes, cutting through the snow, and then would drop back down the file to the end, the easiest position, where the pulk seemed to effortlessly glide along the snow and you could really zone out, just staring at the pulk in front of you, thinking about nothing, almost as if you were meditating. I would get into a comfortable rhythm with the pace being dictated by someone else, much in the same way as track cyclists do – there was little to worry about, apart from putting one ski in front of the other and then, as the hours ticked by, you would gradually work your way back to the lead position and the agony would begin all over again.

On the other hand, there were advantages. A solo trip meant that the only person I had to worry about was myself. I wasn't burdened by the unending pressures of leadership. If I felt that the going was good and I wanted to go further or faster, I could do so. I could change my plan to suit me, and that came with a very powerful sense of freedom and destiny. On SPEAR I could only increase the day if everyone else was in agreement, so we stuck to an established routine. On my solo expedition, however, every day began afresh, and that was truly liberating – until the dreaded whiteout hit, when you become incredibly dislocated from your surroundings. For the solo polar traveller, an hour of whiteout is mentally exhausting – twelve hours and you feel as though you are losing your sanity. Which is exactly what happened on Day 5.

The day began well, with another perfect clear blue sky, little wind and a temperature cool enough to prevent overheating. My spirits were high as I crawled out of my tent for my morning ablution, which – as always – would be done as swiftly as possible. I would dash out of the tent in my tent boots, thermal leggings and down jacket and, with my little carbon-fibre shovel, affectionately known as 'Dug', dig a hole and get it done. Then, instead of using toilet paper, I'd use the chunks of snow I'd dug from the hole to clean up, which was a bit bracing to start with but I soon got used to it.

Having finished my business, I noticed that Colin had camped around 500 yards behind me. He was already up and packing away his pulk. I was impressed. I had gone to bed the previous evening with him nowhere in sight, which meant that he must have skied a long and hard day to catch up. His technique might not have been up to much, but his determination was very impressive. I suppose it was then and there, staring at Colin's tent across the vast expanse of ice and snow, that I realized he was extremely serious about the race. Colin wasn't just there to complete the traverse, he wanted to be the first person to do it. About an hour later, as I was dismantling my tent, Colin skied past me about 50 yards off to a flank. I shouted 'Morning' and he gave me a wave.

Colin was about half an hour ahead of me, so I packed away my equipment and began skiing off in the same direction, due south. I wasn't racing, but I slowly caught up with Colin just as whiteout began to set in. The light went flat and I lost the horizon and almost immediately began to slow. Gradually I felt my morale begin to slip and I grew anxious, wondering how long the whiteout would last. Definitely hours, by the looks of things, but hopefully not days, I said to myself. I pressed on, head down, staring at the compass and occasionally stumbling, which I knew by now was inevitable. I managed to catch up with Colin within a couple of hours and we were soon skiing in parallel just 20 yards or so apart. The white darkness that had descended upon us had closed out the rest of the world, and it felt as though we were the only two people on the continent. Although he was just a few yards away he would occasionally disappear, only to reappear a few minutes later like a ghostly apparition.

As the day wore on, I thought it might be nice for us to have a chat. I glided across the ice – so that at the same time as skiing forward I was moving sideways – and drew closer to Colin.

'How are you doing?' I shouted cheerily. 'It's been handy having you as a reference point up ahead in this whiteout, I'm happy to return the favour if I end up in front.' There then followed a brief discussion, where I mooted the idea that if we happened to still be neck and neck in two months' time on the far side of the continent, it would be incredible to ski over the finish line together. I'd seen it done before when the Norwegian skier Aleksander Gamme, who was conducting a solo, unsupported return trip to the South Pole, stopped short of the finish and waited for two Australian skiers who were attempting the exact same journey. All three crossed the finish line together to go into the history books, and for me it was the ultimate gentlemen's sporting gesture.

Colin then stopped in his tracks. With a sombre face edged with slight frustration, he turned to me, lowering his face mask, and said, 'I want to pass you all the love in the world, Lou, I really do. But I don't think we should speak. I think we need to preserve the solo aspect of this journey. I came here to do a solo expedition, and it's important to me that I achieve that. I hope you understand.'

I was slightly surprised because I had no intention of chatting to him for the next thousand miles or so, but I also respected his wishes.

'Yes, of course, Colin, that's absolutely fine.'

I pulled off and away from Colin until we were about half a mile apart, and then continued in the same direction. Throughout the day we swapped places. One minute I was ahead and then I would lose sight of Colin in the whiteout, and then an hour or so later when the light improved, I would see him powering away and occasionally nervously glancing back at me – it was a sort of game of cat and mouse, which actually I enjoyed, because by the time I felt ready to stop at around 7 p.m. we had skied 15.4 nautical miles, the longest daily mileage of the expedition up until that point. Colin, meanwhile, disappeared into the whiteout.

While I was happy to have notched up a significant number of miles, I refused to allow myself to fall into the trap of racing. These were the early days of the expedition and we were still feeling fit and strong, and I knew from experience what was coming. Colin eventually finished at around 8 p.m. As the whiteout disappeared as quickly as it had arrived, I saw him making camp in the distance.

I woke at around 7 a.m. the following morning and left my tent for my morning constitutional. It was a crisp, clear, beautiful day, and as I scanned my surroundings, I noticed that Colin was already on the move and was little more than a speck in the distance. He must have been up and on the move really early, and clearly wanted to put some distance between the two of us. I was determined to stick to my plan and not to get drawn into racing him just because it made for a good newspaper story. If he wanted to charge off ahead – good luck to him. Besides, I wanted to enjoy the expedition, I wanted to connect with Antarctica in the way that I had done on my previous trips, and I was not going to be able to do that if my main focus every day was pushing as hard as I could.

The climb to the South Pole was sometimes mystifying because I often found myself going downhill before climbing up a huge snow bank, sometimes several hundred yards in size. The steepness of some of the banks would often force me to traverse up while manhandling my pulk over lumps of sastrugi. It was slow, exhausting work.

Where the snow flattened out, I was beginning to notice that the sastrugi was also slowly becoming larger, presenting more of an obstacle. In that early part of the trip, it was just a few inches high, but those strange ice formations were an indicator of what was to come. As the day wore on, the visibility would improve and I could see Colin up ahead, and then the weather would close in and he would disappear, almost as if he had never existed. I found I preferred it that way. I liked the idea that I was alone, just me, the snow and the sky.

By the end of Day 7, I had notched up another 14 nautical miles and had reached an altitude of 2,500 feet. Every day until I reached the Pole, I would be climbing around 200 to 300 feet, until I reached around 9,000 feet. From then on it would level out before a steep descent down the Leverett Glacier. The pulk would be lighter and the going, hopefully, would be easier. But, for now, all that mattered was putting one ski in front of the other. I felt good and well prepared for what the next few hundred miles had to throw at me – or so I thought.

I set off early the following morning, at around 8 a.m., keen to get some miles under my belt, but what I had hoped would be a relatively straightforward day soon became a real battle. The wind began to build and I was soon skiing into a 25 mph headwind. The temperature had dropped to below -25°C. The high wind and low temperature meant I had to wrap up carefully to avoid cold damage to my face. Minor cold injuries were an inevitable consequence of long-distance polar travel, but it was important that I kept them at bay for as long as possible.

I pulled the long, wire-rimmed, tube-like hood of my windproof jacket over my head, which provided some protection against the biting wind. With my goggles fixed in place and my face mask protecting my skin, I felt cocooned against the elements, but it was my least favourite way to travel. My field of vision had been reduced to a small aperture, which affected my balance and left me unsteady on my feet. I could hear myself breathing and my pulse banging inside my head, and there were times when I wondered whether I could actually hear my own heartbeat. Within an hour or so a huge, frosted icicle was dangling from the fur ruff around my hood as the water vapour in my breath condensed and then immediately froze.

The wind kicked up and the spindrift began to move like sand in a desert, with the dusty snow snaking around my feet. The effect was mesmerizing, as if the entire continent beneath my feet was moving. The spindrift built up in front of the pulk and acted like an anchor slowing my progress, sometimes to a halt. It was, I thought at the time, just the Great White Queen looking for a chink in my mental armour and letting me know who was in charge. Day 8 was also the moment I entered my other world, where only I exist. Slowly, thoughts of my family, my home life, a night in front of the TV with a glass of wine were being locked out. They would do me no good on this journey, I told myself.

The only thing that mattered every day was progress, the weather and the surface conditions.

The days began to blend into one another. The hours would tick by, measured by the number of times I stopped for my grazing bag breaks. I would sometimes slip into what was almost a trance-like state, especially in the afternoon when my energy levels were low. My eyes never closed but I certainly drifted away mentally. Whole chunks of time would pass by unnoticed, especially when the going got a little easier or the wind abated.

The best moment of every day was when I chose to stop and make camp. I would find a good flat spot, take eleven more steps, then ditch my harness and skis. It sometimes took all my effort not to simply collapse on my pulk and just sit resting and contemplating for a few minutes. But that was not my routine. The moment you stop moving in Antarctica, body heat is quickly lost. As soon as the pulk harness slipped from my aching shoulders I would begin erecting the tent and making sure it was secure against anything Antarctica could throw at me. I would throw my bedding pack inside and then my food and tent bag. Once in the tent for the night, I avoided going outside again unnecessarily. I even mastered the dark art of peeing inside my sleeping bag by carefully rolling onto my side and positioning my hospital-style pee bottle in the right spot. It was a matter of personal pride that I never had a single spillage.

It was during this period that the worsening conditions caused my daily mileage to drop. I was hit by a triple whammy of increasingly difficult terrain caused by sastrugi, high winds, and poor visibility caused by whiteout. On some days I would spend up to twelve hours skiing but would cover less than 11 miles.

Expeditions like this were always a combination of great excitement and endless repetition, with the odd dose of nerve-jangling jeopardy thrown in. There were hours and sometimes whole days where my senses would be numbed by the monotony of endless white, and I would pray for something – anything – different to break up the day. On Day 10, something did just that, injecting some excitement into my life for a few brief minutes.

As I skied along, picking my way through an area of really awkward sastrugi, I saw something moving in the snow. I stopped dead in my tracks and wondered whether I was imagining things. It crossed my mind that it might be some form of wildlife, but quickly reminded myself that this was just too unlikely. Could it be a stone, or a rock? Whatever I had seen was constantly appearing and disappearing as I moved up and down amongst the undulating terrain.

Antarctica can play tricks with your eyes, much like the desert, and estimating distance was extremely difficult because there were few reference points. I couldn't tell whether I was looking at a large object a long way away, or a small object that was relatively close. I was sort of spooked, if I'm honest, but I kept on skiing for around 300 yards until I reached the lump of sastrugi where this thing was. I gingerly inched myself closer and closer until I reached the fluttering object. As I bent down to examine it, I saw it was not some strange Antarctic creature, but a small plastic sticker on which had been printed the stars and stripes. Colin's ski tracks led away from it. I realized that the flag must have come unstuck from his pulk. Laughing to myself, I decided to keep it as a souvenir, and to remind him – if and when we ever met up again – that littering in Antarctica was prohibited.

I made camp that night with two things on my mind. Firstly, the joy of rolling up a ten-day food bag, knowing that my pulk was 15 kilograms lighter. I had seven of those large bags; at the end of each one, I knew I was ten days closer to achieving my goal. And secondly, it was time for a chocolate-chip biscuit pudding. It was a sumptuous dessert packed full of calories. I slept well that night with my mind focused on emptying the next ten-day food bag and enjoying another chocolate pudding. Such is the life of an Antarctic traveller: it's all about the little things.

In fact, Day 11 began with the memory of my ten-day chocolate pudding still at the forefront of my mind. I was torn between enjoying how great it tasted and feeling sad that I had to wait another ten long, hard days before I got to the next one. Setting off on the ice with a set of fresh socks was also bliss. But very quickly I began to feel fatigued, and the day turned into a real struggle. I was completely caught out by how difficult I was finding hauling the pulk through the sastrugi.

By the end of the day I was completely exhausted, my legs like jelly. I struggled to remember ever feeling so exhausted, even during SAS selection. I could barely walk, and I crawled around while putting up the tent and digging in the valance. Preparing my food was a real struggle when all I wanted to do was sleep. I had hoped that the Thiel Mountains would have made an appearance by that stage, but they were still several degrees away. I tried to rationalize why I was feeling so tired and could only assume that I had hit a bit of a wall. I estimated I had already lost around two kilograms in weight and I had been on the move for eleven days nonstop. Had that been a mistake? My plan had been to only take rest days when the weather was too bad to ski. It was a sound strategy, providing that the rest days weren't too far apart.

Day 11 had been hard but Day 12 was even harder. I wrote in my diary:

Double whammy. Whiteout and sastrugi. Absolute nails. Stumbling around blindfolded on an icy obstacle course dragging what feels like a bathtub with a fat bloke in through thick custard.

It was a demanding day that offered up just 10.7 nautical miles in ten hours. But experience had already taught me that I should not fight Antarctica. Bad days would come, and I knew there would be many more. I had to work through each day and, as Shackleton had said, I needed to let Antarctica into my heart.

I woke rested on Day 13 to clear blue skies and light winds. The sastrugi was still an unrelenting challenge, but I took a lot of comfort and pride from not getting angry and moving through the ice fields with a Zen-like calm. My mileage was again low, but my morale was high and that was important.

Then I made the mistake of thinking I was on top of things and had it all under control. It was a basic error. On Day 14, two weeks in, I suddenly cracked. The sastrugi was making skiing almost impossible for anything further than a few feet. I was effectively man-hauling a 110-kilogram weight across the ice and it was slowly breaking me. Every third or fourth step I seemed to stumble and fall, and my energy levels had fallen off a cliff. I had pushed myself too far without taking a break, and now I was paying the price. On the previous expedition we had taken a rest day every couple of weeks, irrespective of the weather. All I thought about throughout the day was making it to the evening, but as I skied forward, falling over the sastrugi again and again, my willpower – the one personal quality that had always proved so reliable and had got me through so much in life – began to fail. At one point during the afternoon I stopped skiing, walked back to my pulk and collapsed in a heap on top of it. I could barely lift my head from my chest for half an hour, thinking: this is so hard. I was close to breaking down. I had entered the unknown and I didn't know how to respond. There had been tough days on my two previous expeditions in Antarctica, but this was something else, and the daunting reality of knowing I had at least six more weeks of hard pulling and skiing ahead of me was almost too much.

I struggled to my feet and somehow managed to ski for the rest of the day, but by the time evening came I was barely moving. I used the last of my willpower to erect the tent before collapsing inside. I felt like a broken man; at that moment in time I honestly believed that my expedition was over. I had entered a dark, lonely place, devoid of answers or self-belief. Never before in my entire military career had I felt like this. For the first time I had accepted failure and defeat; in fact I welcomed it. I just wanted the expedition to be over and to go home. If I'd had the strength, I think I would have broken down into a sobbing mess, but the truth was I just felt empty, as if I had exhausted every last ounce of energy in my body.

I took off my boots and slowly began the process of preparing my meals, trying to find reasons why I should carry on. I knew in my heart that life was just going to get harder and I would become weaker. The limited amount of body fat I had arrived with two weeks ago was now long gone. My energy output had been exceeding my calorific intake for the past two weeks, and I was paying the price. The solo traverse was not something that could be achieved by a fifty-year-old man, I told myself. I decided that I would eat my meal, then contact ALE and tell them that I was giving up, that I had failed. I decided that I could live with the shame. Everyone has a breaking point, and I had reached mine. There was nothing more to be done.

Then, out of nowhere, I stopped and paused. As if a voice at the back of my mind was saying 'just hang on for a minute'. I had at least forty more days to go, six weeks of suffering, hauling and falling. Six weeks of rehydrated food and frostbite and blisters. That was one option. The other was a lifetime of regret. With luck I might live to ninety. I was faced with the choice of putting myself through a living hell for the next forty days and being able to sit back for the rest of my life in the knowledge that I had achieved something historic; or I could throw my hand in and face the possibility of regretting my actions for the next forty years. I had seen at first-hand how people could be negatively affected by giving up in Antarctica, and I didn't want to become one of those casualties.

It was a case of me turning to myself and saying: 'For God's sake, Rudd, man up.'

I sat in my tent and tried to rebalance myself, feeling confused at how far I had slipped. Slowly, I came to the conclusion that I was physically and mentally exhausted, almost incapable of rational thought. I needed to rest for maybe twenty-four hours and then make a decision. I'm still not quite sure why, but I began videoing myself. In a wide-eyed rage, I told the camera I wanted to block out the outside world, civilization, my family and ALE. The only way I was going to be able to complete the journey was to be at one with Antarctica.

I spoke about how much I loathed having to call in to ALE and having to speak to a human being and give them a sitrep of my progress every evening. I hated wearing a wristwatch and having any concept of time, because it didn't matter. It was irrelevant. I decided that I wanted to rid myself of all things technological. It was annoying to have to carry a satphone, a tracking beacon and having safety cover, because it didn't feel like a truly immersive journey. Neither Scott nor Shackleton had any of these devices, I ranted to camera. I came to the conclusion that the only way I could complete the journey was to rid myself of all signs of civilization and completely isolate myself from the rest of the planet. In a mad frenzy I decided I wasn't going to listen to any more audiobooks or music, and I cut up my headphones with a knife. I was losing it.

I was seconds away from phoning Wendy and telling her that I wasn't going to do any more blogs, when I paused and thought about my next move. Wendy had the power of abort over the expedition and, if I had told her what I was thinking, she might have decided I was unravelling. So, when I became a bit more rational, I realized that phoning Wendy was not such a good idea. My tirade lasted for another thirty minutes, and then I calmed down and immediately regretted cutting up my headphones. Luckily,

I had a spare set. Needless to say, the video has remained a secret, unseen by anyone.

The way forward, I decided, the way I could keep going and retain my sanity, was to remove the pain and frustration from my mind. Think of nothing but the next step. I likened myself to a donkey I had seen in Afghanistan, loaded with weight and just putting one foot in front of the other. That is what I had to become; I had to not worry about anything else. I slept like ten men that night, waking early and feeling refreshed and relieved that I had come to the conclusion that I wanted to continue.

But as I climbed out of the tent to begin packing away my equipment, I felt my heart sink. Although visibility was good, a headwind was gusting at around 30 mph and the temperature had dropped to around -30°C. It was going to be another tough day. Around an hour later, gear packed away, I set off in full kit – goggles, face mask, hood up, and mitts rather than gloves. The mitts kept your hands warmer, but they were a real pain. Zips, buttons, anything that required a bit of manual dexterity were a nightmare with mitts.

The wind got stronger and stronger throughout the day. That evening I managed to get the tent up without too much difficulty, but it took almost an hour. Throughout the night the wind howled and turned into a full-on gale, and I slept little, almost like a cat with one eye open, waiting for the tent to collapse.

I woke in the morning hoping that the wind speed had subsided, but within seconds realized that, if anything, it was now stronger. I stuck my head outside and saw that it was too strong to safely take the tent down by myself. If I lost the tent, it would be game over. A single gust, a moment's lapse in concentration when taking down a tent in a gale and it will sail away. There was no other option but to remain inside and wait for the wind to ease.

By 3 p.m. that afternoon the wind had dropped just enough to allow me to get the tent down and begin skiing. Initially the wind was still quite strong, and I now faced the prospect of having to ski until the wind speed dropped significantly: if it remained high, it would have taken me over an hour to get the tent up, during which time I would run the risk of getting exposure. Fortunately for me, by 8 p.m. that evening the wind had abated.

As can only happen in Antarctica, sheer misery can be replaced by joy in a matter of moments. As I skied southwards, the snow-capped peaks of the majestic Thiel Mountains emerged reassuringly on the horizon. Immediately I felt my spirits lift. I knew the mountains would be coming into view, but their welcoming presence was like meeting an old friend.

ALE have an unmanned ski runway at the base of the mountains, along with a couple of Portakabins and vehicles. They use the site as a resupply point for some expeditions, as well as a refuelling stop for aircraft. I deliberately stayed around half a mile away to avoid any contact just in case there was anyone stationed there. I moved on quickly, and on the horizon noticed that the sky was darkening. I sensed that more bad weather was heading my way.

Within a few hours it began to snow, lightly at first and then much more heavily, accompanied by a driving wind. A lot of the time when you experience snowfall in Antarctica, it's actually snow that's been picked up by the wind from another part of the continent and then deposited elsewhere. It snowed on and off over the next two days and, by Day 19, around nine inches of soft, powdery snow was lying, and having a severe impact on my progress. With each step the snow would build up in front of the pulk, acting like a brake. Pulks generally have two runners and are designed to glide along on a firm surface. The friction created by the runner moving across the ice melts a thin film of water, which acts as a lubricant. However, experience had taught me that this only works within a certain temperature range and with the right surface conditions. Too cold, and the melting doesn't occur, and the pulk almost sticks to the surface; too warm, like I was experiencing now, and the pulk sinks beyond the runners and drags along on its belly.

I was really struggling to make any meaningful progress, managing just 100 yards or so before I would have to stop, bent double over my ski poles, gasping for breath. It was time for a rethink. I decided to remove half of the weight from my pulk and cache it in the snow. I marked the location on my GPS so that I had an exact fix, and then skied forward for a couple of miles with the other half. I then unloaded the pulk, marked that location with my GPS, then skied back with an empty pulk, picked up the kit I had stashed and skied forward to where the remainder of my equipment was located. It was a long, slow and very boring process, but it was the only way I was going to make reasonable progress, given the soft snow conditions. In effect this meant that to move forward one mile I had to ski three miles: pretty demoralizing work. By the end of the day I had skied for eleven hours but covered just six nautical miles.

I had never seen so much snow fall in such a short period of time in any of my previous expeditions, and I did wonder whether this was a weather phenomenon linked to climate change. The staff at ALE, however, were a bit more circumspect, and said that they believed a bad summer season came around every four years and this was the fourth year. I was subsequently to find out that at least four other soloists that season, who came through those same conditions, all aborted their expeditions to reach the South Pole. Among them was the highly experienced Eric Larsen, who was attempting a speed record, and Jenny Davis, a professional athlete from the UK.

I had hoped for better conditions the following day, but the snow was just as soft and I realized immediately that I would have to adopt the same strategy as on the previous day. But my progress was even slower. The snow was so soft that even skiing was hard work, but the visibility was good so I tried to focus on that small positive. Unfortunately, I became a bit too blasé about

GPS-marking my cache site, and I fell victim to the dictum, 'familiarity breeds contempt'.

The good visibility and the flat terrain meant that I could just about see my cached equipment – which was marked by a spare ski – when I reached the turnaround point, so to save a bit of time I stopped the process of GPS-marking, which took around five minutes. My GPS was tucked away inside layers of clothing and turned off so that I didn't drain the battery. To switch it on meant I had to dig it out, fire it up, wait for it to get a fix and mark the location. It had become an annoying delay each time in an already time-consuming process. Besides, each time I turned around to ski back, I was following a virtual motorway of my own tracks. What could possibly go wrong, I thought?

This worked well until around the middle of the afternoon when I finished one run, turned around and noticed that the weather had changed. I was quickly enveloped in whiteout, the wind picked up, and within a few minutes my ski tracks had all but disappeared. It was a basic mistake and – not for the first time – I felt a mild surge of fear. My main concern was that I had left behind a few food bags, together with my tent and sleeping bag. Normally I would always have one of those – so if I lost the tent I would always have a sleeping bag and vice versa. I had made a potentially life-threatening mistake and I did my best to control my fear. What was needed was a clear head and rational thought if I was going to successfully find the cache.

I began skiing back in what I hoped was exactly the same direction I had come, using the compass on a back bearing. I searched for my tracks in the snow but, as the wind picked up, they quickly became filled with spindrift and were soon disappearing. Even getting down on my knees to try to identify them in the flat light proved fruitless.

I knew that, given the diminishing visibility, I would have to get within 30 feet of my kit if I was to have any chance of spotting it. If not, I would be reduced to digging a snow hole, putting on all my warming clothing and calling in an emergency re-supply or pick-up from ALE.

I was now skiing slowly on a compass bearing, pacing and scouring the snow as I inched my way back to where I hoped my kit was located. I had skied forward for an hour and a half, so now I had to do the reverse and try to ski back to the exact start point. I was constantly scanning left and right in the hope of catching a glimpse of my gear or a flash of colour against the encroaching whiteness. For an hour and a half, my heart was in my mouth. Frankly, I was absolutely terrified. As I skied on, I became convinced that I had missed my kit and so I skied back on a parallel track. At one point I took off one ski, stuck it in the snow and then walked off to a flank, constantly checking that I could still see the ski. I then returned and did the same on the other side. I was using a box-search method that I had employed many times on operations, whether that be searching for IRA weapon caches in Northern Ireland, or clearing ground of IEDs in Afghanistan.

Just when I was about to give up, through a fleeting gap in the whiteout I spotted my upright spare ski. Thank God, I said out loud, and offered up a quick prayer of thanks. I skied over and was almost overcome with relief when I found my equipment slowly being consumed by spindrift. In another hour or so it would have completely disappeared, with just the spare ski I'd left behind poking out. I dropped to my knees and vowed never to make such a stupid mistake again. It was a schoolboy error that could have ended the expedition or – even worse – put me in a life-threatening situation. The one positive of that day was that it was Day 20, which meant a clean pair of socks and a chocolate pudding. There's always a silver lining if you look hard enough.

That night I called home for a prearranged chat with Lucy, and briefly told her about the day's exploits. It was good to hear her voice and reconnect with my other life, despite what I had thought a few days earlier. I then watched *The Darkest Hour*, the story of how Winston Churchill turned the tide of Nazi oppression during the Second World War. I needed a bit of inspiration after a day where I had skied for eight hours but only progressed three miles. As Churchill said, 'If you are going through hell, keep going.'

16

SNOW PETREL

Life is either a daring adventure, or nothing.

HELEN KELLER

I began Day 21 chastened by the previous day's mistakes, and determined never to repeat them. I had gambled and almost come unstuck. The lesson was well and truly learnt.

The snow was just as bad as the previous couple of days, and for the first few hours it felt as if I was trying to drag the pulk through treacle. But just as I was beginning to despair, the surface began to improve, probably because I was moving up a small slope; you find that as the wind accelerates down these inclines, it helps to harden and compact the surface. As I pushed forward, I noticed that little mounds of ice a few inches high began to appear. In between these mounds were troughs where the deep, soft snow had gathered, so I ended up bounding from one mound to the other. I got into a routine of getting on top of one patch of hard ice, searching for another and then charging at it through the deep powder. I would then stop, out of breath, legs screaming with lactic acid, and look for the next mound. I wasn't making much progress, but it was the only way of coping with the soft snow. In the latter half of the day it improved, and I was once again on terra firma. Then, as had happened so many times before, when you think you are making a bit of headway, Antarctica reminds you who's the boss.

By around 3 p.m. the wind picked up. It got stronger and

stronger throughout the rest of the day. It was a katabatic wind that was tumbling down off the higher Antarctic plateau, gathering speed as it moved across the flat ice, and it was hitting me head-on. I was now faced with another difficult judgement call. Should I stop and put my tent up, or ski on? If I decided to ski on and make up time I had lost over the last few days, I might face a situation where the wind strength had become too strong for me to safely erect the tent. High winds can last for days, and I could end up skiing on for hour after hour until the wind dropped sufficiently. Despite the obvious risks, I chose the second option, gambling that the wind speed would remain the same or even ease. But it didn't. Within a few hours it had reached 45 mph. Once again, I was left questioning my decision-making. My regimental motto might have been 'Who Dares Wins', but I think I was pushing that to the limit.

Thankfully I began to notice a rhythm to the wind. It would howl like a beast, buffeting one way then another, punching me in the chest and forcing me backwards, but then drop into a lull that would last for around ten minutes. It would then build and build until it was gusting back at 45 mph. I monitored the conditions over the next few hours and began to hatch a plan. If I put in a really good shift, I reckoned that I could ski for twelve hours and then pitch my tent and make camp during one of the lulls.

I pushed on throughout the day, refuelling myself from my grazing bag every hour or so until around 8 p.m., when I waited for the lull. When it came I went into overdrive, working as quickly as possible – but without panicking – while mentally counting down the minutes until I knew the violent gusts would return. It was all going really well and as predicted, until suddenly a powerful gust came out of nowhere and sent the guy lines into a spinning tangle. The pegs were always left attached to make pitching quicker, but when they were caught by the gust they became entangled like a bag of snakes. I stared in horror at the mess and felt the wind speed beginning to build again.

Forced to remove my gloves, I begin picking away at the mass of ropes, which now looked like a badly made bird's nest. I had to be careful not to touch the alloy snow pegs; at those temperatures my fingers would have stuck fast to them. I lay flat on top of my tent, spreadeagled to prevent any of the edges getting caught by the wind. Eventually I managed to pull each of the guy lines free and get the tent up before the wind reached its full strength once more.

The wind continued throughout the night, building and dropping, and I slept badly, waking with a start every time the tent began to buckle and bend in the strong gusts. In the morning I woke tired, but relieved that the gusts seemed to have dropped away completely.

The strong winds over the previous couple of days had started to harden the surface – just in time, as the next obstacle ahead of me was a steep climb up a huge, rolling snow bank. It was hard graft from the off, and there was nothing I could do but grit it out and push on, head down, puffing, panting, and trying not to overheat. The day was made harder by a series of almost endless false summits. The pulk felt as though it had gained another 50 kilograms and I was forced to almost drag it up at times. But by lunchtime I was clear of the climb and skiing on relatively flat snow and ice. The wind dropped and I made some really good progress until around 7 p.m. when I stopped and made camp. It had been a tough start and I was very tired, but my spirits were high, almost as if I had been tested by Antarctica and come through it in one piece. My thighs were killing me and once again the skin had been stripped from my heels, but I felt in a good place. As the saying goes, there's no such thing as an 'easy' day in Antarctica, they are just different.

As I went to bed that night, I decided that my focus must now be hitting the degree lines until I reached the Pole. Degrees of latitude are 60 nautical miles apart, and I had four more degrees to go to reach 90 degrees south. The last few days had been one of the toughest periods of my life, but I had somehow prevailed and my confidence was building. I was incrementally expanding what I thought I was capable of, and it was giving me a different perspective of my perceived limits.

Day 23 was kind to me. It was good weather all day and the temperature was a manageable -25°C. It was, in many respects, a perfect day. Too warm and the pulk tended to sink in the snow, but -25°C was just right. I managed to ski for eleven hours and reach an altitude of 5,251 feet, with each step taking me closer to the Pole and further up onto the polar plateau. The weather gods had gifted me a few hours of perfection. It was quiet, almost windless, and as I skied along I heard the unmistakable drone of an aircraft flying high above. I gazed upwards and began searching, and there it was, heading who knows where, marked only by a vapour trail across a faultless aqua-blue sky. I stopped and stared for a few moments, almost tempted to wave, and wondered what those on board were doing. Were they awake, sipping on champagne, looking down at the vast white continent beneath them or tuned in to an in-flight film. They were in another world, completely different to mine, but I wasn't envious. I could travel on an aircraft whenever I wanted, but I might never get an opportunity to repeat a journey like this.

Over the next few days the weather flip-flopped between whiteout and great visibility. I felt in control and was routinely knocking out between 13 and 14 nautical miles, until 29 November, Day 27, when I found myself skiing in some monstrous sastrugi in whiteout conditions. I had set off in fight mode, crossing 86 degrees south and heading towards 87 degrees south. I had already skied for around 400 nautical miles and was absolutely determined not to be deflected from racking up another decent day's worth of mileage. But the sastrugi were like nothing I had encountered before. They weren't just small ridges in the snow, but huge boulders that had been sculpted by winter winds. It was like walking through a frozen maze with almost no visibility.

It was sastrugi hell, and within a couple of hours I had fallen over around twenty times. At one point I unknowingly climbed up onto a large piece of ice, which was around 300 square feet and eight foot high, and was skiing along thinking that I was at ground level and feeling happy that I had reached an area that was relatively clear of sastrugi. I went for a head-down burst. All was going well until I came to a sheer mini cliff edge, which I couldn't see, and skied straight off the end into a void. As I felt the ice disappear from beneath my skis and I went into freefall, I thought I had fallen into a crevasse. In that split second I thought I was dead, that I would hit the bottom at some point and would ultimately die from the cold if I wasn't killed by the impact. But less than a split second later I crash-landed face first into ice as hard as rock. The metallic taste of blood filled my mouth and I was seriously winded – my diaphragm had gone into spasm and was preventing me from breathing properly.

I blearily saw the tip of my ski split open and come apart like a banana skin when I hit the ice, then a split second later the pulk landed square across my back. It was like being hit by a double combination thrown by a heavyweight boxer. The pain surged up through my body, so intense that I was unable to move for at least a minute.

I assumed that I had done some serious damage. The pain in my legs was so bad I thought I had broken one, if not both. Gradually I managed to move my fingers, then my arms and then my legs. I slowly shuffled out from beneath the pulk, expecting at some stage to see a bone sticking out of my thigh or some other horrendous injury. Then gradually, as I came to my senses, it dawned on me that I had been incredibly fortunate. Somewhere, I thought, my guardian angel was on watch. I was battered and bruised but – other than a bit of damaged pride and a broken ski tip – I had survived pretty much intact. I sat by my pulk for a few minutes, just waiting for the adrenaline to ebb away.

I was beginning to get cold so I decided I had two choices. Put the tent up amongst that sastrugi and wait for the weather to clear, or cautiously press on. I chose the latter, assuming that it was better to get back on the horse straight away. I stood up and began to gingerly ski forwards, trying at the same time to feel the ice with my skis. The last thing I wanted was another heavy fall and so, for the next thirty minutes or so, I sort of half walked, half skied until my confidence began to return.

I stumbled and fell throughout the rest of the day and assumed that I was hardly making any progress, but that night as I checked my GPS I found to my amazement that I had managed to ski just over 13 nautical miles. I also managed to effect some sort of basic repair on the ski, but I was going to have to use my spare set from that point onwards. As I went to bed that night, I came to the conclusion that someone was looking out for me. In the last few days I had suffered a series of potentially expedition-ending incidents and, although I hadn't escaped entirely unscathed, I was still very much in the game.

Day 28 was another punishing day of whiteouts and sastrugi, and I began to wonder how much more I could take. My knees and hips were badly bruised and I was in almost constant pain. It felt as if I was falling over and crashing into the ice with every other step. The ice in Antarctic is as hard as rock, and the constant impact was beginning to take a severe toll. I now understood why sastrugi had caused so many previous expeditions to fail. I felt completely drained of energy, and the effect of whiteout turned the experience into a waking nightmare. There were times when doubt began to creep back into my consciousness. It became a battle of wills: mine against Antarctica's.

Throughout that day I repeated the opening lines of the Dylan Thomas poem, 'Do Not Go Gentle into that Good Night'. It became my mantra. I was in attack mode, and I was going to get to 87 degrees south, regardless of whiteout, regardless of sastrugi. It had been one of the longest and hardest days on the ice and I was mentally exhausted. I spent thirteen hours smashing my way through the sastrugi, and finally stopped at 9 p.m. I was now 180 nautical miles from the Pole, and my next target was the two fat ladies – 88 degrees.

By Day 30 I started to wonder whether the sastrugi field I was trying to ski through was going to stretch all the way to the South Pole. The difficult stuff had disappeared, but it was still something of an obstacle course, and I was just praying for a few miles of relatively flat going. It was a cold, windy day, and I was marching head down in full polar gear, hood up and goggles on, when I noticed some movement out of the corner of my eye. Up until that point I had been cocooned in my own little world, so this sudden change startled me.

I stopped in my tracks and looked up, pulled my hood down and lifted my goggles, and there, right in front of me, was the most stunning pure white bird, around the size of a dove with a jet black beak. Initially I wondered if I was hallucinating. I was very close to the centre of Antarctica. Nothing lives there. It is the most inhospitable place on earth. But there it was, just 15 feet away, flying in front of me at head height, almost staring directly at my face.

A crosswind was blowing hard and I realized the small bird must have been riding the air current across the continent when it decided to check me out. I didn't know it at the time, but I was having a close encounter with a snow petrel. It was clearly just as surprised by the meeting as I was, and began to circle, scrutinizing me intently. For all I know I could have been the first human it had ever come across. I stood there in open-mouthed amazement. I had never seen or heard of any wildlife this far into the interior, and thought that I should get a picture of the bird so that when I retold the tale later, people wouldn't think that I had gone nuts. I shuffled back to the pulk to get my grazing bag and camera and, just as I scattered some nuts and dried fruit onto the ice, the snow petrel was buffeted downwind. By the time I had managed to get my phone from inside my jacket, the bird was gone.

The whole event had only lasted at most for a couple of minutes, but it had a profound, almost spiritual effect upon me. I sat on my pulk convinced that the bird was some sort of visitor. It was Henry's spirit, I said to myself; he was checking up on me and wanted to let me know that he was watching over me. I know it sounds odd, but the experience was just too surreal and I still believe that to this day. Just to be clear, I am usually a total sceptic as far as ghosts and spiritual events are concerned. But the chances of that encounter happening were so rare that I was 100 per cent convinced that it was a sign. Moreover, I was also absolutely convinced that I was going to complete the journey. That day was a turning point in my mental approach, and from that moment on I never doubted myself. I realized that when I spoke about the encounter on the audio blog, people were going to think that I was losing the plot, and there were a few friends who joked that I was hallucinating, but I wasn't, of that I'm sure.

The good weather continued for the next couple of days, and on Day 31 the crowning glory was a huge 'sun dog', a halo that forms around the sun. Its scientific name is a 'parhelion' and the phenomenon is caused by the refraction of sunlight by ice crystals in the atmosphere. Sun dogs tend to occur in the Arctic and Antarctic regions when the temperature drops to an icy -20°C. There have been reports of as many as four or five rings forming around the sun when the conditions are right, and they always make for an amazing sight. I got into a routine of saying 'hello, son' every time a parhelion made an appearance, as a nod to my son Luke, who was back home serving in the Marines. I had named my compass after my youngest daughter Sophie, a criminal barrister, for always guiding me through the whiteouts; my skis after my eldest daughter Amy, serving with the Royal Air Force, for flying me effortlessly across the surface; and my tent after my wife Lucy, my unfailing haven from the elements at the end of a long, hard day.

I saw the existence of the sun dog as a good omen. I needed a bit of a lift because I was now beginning to feel the effects of altitude, which I had struggled with on previous expeditions. I was skiing at over 8,000 feet and finding breathing increasingly difficult. My head was pounding too. Although 8,000 feet is not particularly high in mountaineering terms, the low pressure zone that sits over the Antarctic has the effect of making the altitude feel more like 15,000 feet. My pace began to slow as the thumping inside my head grew faster, and I made camp at around 6.30 p.m. that evening, hoping that some food and a good night's sleep would help. Instead I woke in the early hours almost in a panic as I struggled to breathe. It felt as if the air had been sucked out of the tent and, with a foggy head and feeling slightly disorientated, I wondered initially if I was in the middle of a dream. Fortunately, the altitude sickness was temporary. I soon adjusted and it became more of a mild inconvenience than anything more serious.

On Day 33, I crossed 88 degrees south. It was a major milestone, and I was now well and truly on the polar plateau. The bulk of the climbing and the hard work was behind me, as was the worst of the sastrugi. I only had to climb another 600 feet and the ice would flatten out; from the Pole onwards, heading north again, the incline would be barely noticeable until finally I would be running downhill. The mere thought of what was to come put a smile on my face as I skied due south. But there were still issues I had to face. The temperature on the plateau could fall to -40°C, and I had a lot less body fat, if any, to help protect me from the cold.

By Day 37, I had crossed 89 degrees south and I was now within striking distance of the Pole. It was an incredible feeling, but I needed to achieve 15 nautical miles a day to crack this last degree in four days. As I skied on, heading due south, I was buzzed by one of ALE's Twin Otters, which flew in low and fast at what seemed like just a few feet above my head, making me jump and laugh out loud. It was further proof that my next goal was close by.

The weather closed in over the next few days and, with limited visibility, I was forced to ski on a compass bearing until I was just 11 miles from my target. By pure coincidence I've somehow ended up camping exactly 11 miles out on the penultimate day of all three of my Antarctic expeditions. It was also Day 40, which meant another food bag was gone, but more importantly I had a chocolate pudding and a change of socks.

The following morning I skied off with a smile fixed to my face as I waited for one of the observatory structures, which form part of the South Pole station, to come into view. Then, through the weak light of the fading whiteout, I spotted it, a vast metal structure almost resembling a rusting shipwreck. It was like receiving an energy boost, and for the first time on the entire trip I felt as if my pulk was weightless. With each mile I completed, more and more of the South Pole station buildings and infrastructure became visible. Eventually I reached a sign that said, 'You are almost at the South Pole Station. From this point please call in.' So I phoned ALE, who provided me with a bearing on which to head in to avoid disrupting the various clean air experiments.

As I approached the two small ALE tents, the staff came out to greet me. The rules are that they can hug me, say hello, take a few photos but not offer any food or drinks. The hugs were a nice welcome, but they are also a subtle way of checking out how much weight I'd lost. They checked me over, looking at my hands, nose and face for signs of frostbite, but I was in pretty good shape given everything. They also told me that Colin had arrived the day before and looked exhausted. He was clearly very tired, the staff said, and had a small amount of frostbite on his nose. Although I had been convinced that I would complete the crossing, I felt pleased to have made it to the Pole unsupported, which was a major achievement in itself; one staff member pointed out that I was one of only very few people who had skied full distance to the Pole three times.

I chatted for thirty minutes or so but soon began to feel the cold and, after a very brief photo session at the Geographical and Ceremonial South Pole markers, I said my goodbyes and pushed on, skiing parallel to the main ice runway and finishing later that evening. It had been an epic day. I had skied for sixteen hours and clocked up 14 nautical miles. The South Pole station was still in view when I went to bed, but importantly it was now behind me and I was heading north. I had 620 miles in the bag and just 300 to go.

17

HOME RUN

It always seems impossible until it's done.

Nelson Mandela

Almost from the day I skied away from the South Pole, my body began to deteriorate. I estimated I had lost more than 10 kilograms in weight and I felt physically weakened. My mind and determination were still strong, but I did begin to wonder whether physically I was capable of completing another 300 miles. I had also developed a badly ulcerated mouth, which made eating and drinking painful and difficult. The ulcers began to form on the day I reached the Pole, and when I woke the following morning, my mouth was a mass of scabs, blood and pus, my lips welded shut by the ooze that had seeped from the mass of sores as I slept. My first action of the day as I lay in my sleeping bag was to gently prise my lips open. I had to ignore the agony, even though my eyes smarted and my mouth felt as if it was on fire.

The ulcers had been caused by the extreme cold that accompanied me as I skied towards the Pole, and I knew that the sores were unlikely to heal until I finished the expedition and left Antarctica. There was also the threat that the ulcers could lead to a wider, more debilitating infection, which could threaten the final phase of the expedition.

Breakfast was a lengthy and traumatic experience, with each mouthful of porridge forcing me to wince with pain. It would have been easy to spend the day in my tent feeling sorry for myself, but the wind had changed direction and there was a good opportunity to get some decent mileage under my belt. I phoned the ALE doctor, my good friend Martin Rhodes, who confirmed what I already knew – I had suffered a cold injury and the best I could do was monitor the sores and try to control the pain with painkillers and steroid cream. Eventually the pain subsided, but the relief was short-lived. As soon as I stepped out of the tent into the cold air it felt as though I had been whacked in the mouth with a hammer. I hadn't realized just how painful cold-damaged lips could be, and I began to appreciate the agony Ollie must have been in during the SPEAR expedition.

By 7.30 a.m. I was on the move with the wind on my back, and I found that I could get a few seconds' respite from the pain if I folded my lips into my mouth. The intensity of the pain left me feeling very tired and made skiing a miserable experience. There were periods when the throbbing disappeared and I felt my spirits begin to rise, and then the pain would appear again, often stopping me in my tracks. At that point I would cover the whole of my face with my hands in a vain attempt to shield my damaged mouth from the icy wind, hoping against hope that the pain would fade away. But it rarely did. I felt as if I was being punished for some unknown crime. I began to notice that the pain had taken on a sort of rhythm, especially when I stopped for my hourly handful of nuts, chocolate and sweets from my grazing bag. Sometimes there was no pain at all, and on other occasions it was intense, even though I was eating the same food. I wondered whether there was something amongst the nuts, seeds, chocolate, sweets and salami that was aggravating the ulcers.

I began to remove one element from my grazing bag at every stop, until I discovered that the intense pain was made much more severe by some fruit pastilles – who would have thought it? There must have been something acidic in the sweet's flavouring or colourings, which was playing havoc with my ulcers.

It had been a long, painful day that left me feeling shattered, but I was now beginning to crest over the edge of the Titan Dome as I climbed towards 10,000 feet. Despite everything, I was happy with my progress. I had skied for twelve hours and managed to cover 16 nautical miles. Now that the incline was beginning to flatten out, the time had come to fit my short mohair racing skins to my skis; they provided less friction and more glide, allowing me to travel much faster. It was a fiddly job, and I had to bring the skis into the tent to warm them up, heat the glue and apply them securely. Well worth the effort, though, as poorly fitted skins will quickly detach in Antarctica and cause unnecessary delays.

The intense pain from my ulcerated mouth caused me to wake early on Day 43, and I was once again on the move by around 7.30 a.m. but I had the wind on my back and the going was good. The pulk had never felt so light and I was soon eating up the miles. The temperature was -28°C and dropping, and I was soon feeling very cold – a pattern that would last for several more days.

As I skied onwards, I had a close encounter with a vast resupply convoy heading for the South Pole Station. It was so large that I was able to spot the vehicles from several miles away, and the ice beneath my feet began to vibrate as they got nearer. There were up to fifteen huge, caterpillar-tracked vehicles pulling massive sledges carrying food and provisions, and huge black rubber bladders containing thousands of litres of fuel.

The supplies were shipped in from the USA and would be offloaded at McMurdo Station on Ross Island onto huge sledges or placed inside containers on skis and attached to the tracked vehicles. From McMurdo the convoys, driven by teams of drivers working shifts, would head towards the Pole, travelling in all weathers for up to two weeks. It was a huge undertaking and not for the faint-hearted. The guys driving the trucks were dressed in checked lumberjack shirts and baseball caps and looked like the tough US truckers that they were. The vehicles didn't stop, but the drivers waved and gave me the thumbs-up as the convoy thundered past. I initially dropped into the tracks left by the vehicles and made some great progress on the compacted surface, but after a few hours they began to fill with soft spindrift and it became a really difficult surface to ski on. In the end I gave up and moved back onto the untouched surface, parallel to their tracks. It made sense to run parallel to their route to reduce the risk of going down a crevasse. The whole route is marked with small bamboo poles around every 400 yards to help the vehicles stay on track, as it would be easy to stray off course in a whiteout. For most of the time I was able to use these to keep my parallel course, but did lose them a few times on whiteout days. By the end of the day I had completed 17 nautical miles, a record for the expedition up until that point.

By Day 44 – four days after leaving the Pole – the conditions were superb. My course had taken me on a dog-leg right turn, and I was heading for the Leverett Glacier – a vast ice highway that would eventually take me down from the Antarctic plateau onto the Ross Ice Shelf. It was another long day on the ice, but by 8 p.m. that evening I had skied 18.5 nautical miles. Despite the pain in my mouth, my spirits were soaring and I felt fantastic. Just as I was about to start making camp that evening, I spotted a tent and some vehicles in the far distance. I considered skiing on to check it out but decided that I had spent enough hours on the move for one day.

I woke early again the next morning and went through the now familiar and agonizing ritual of prising open my infected lips. As I prepared breakfast, I opened the tent door and looked across the ice to the tented camp I had spotted yesterday. Whoever they were, there was little activity, and I assumed as I tucked into my porridge that they were having a lie-in. Lucky them. I was back on the move by 7 a.m. and, as I skied closer to the camp, various people began to emerge. Behind the tents two Hilux four-wheel-drive trucks, which had been specially adapted for travelling across Antarctica, came into view. The trucks could cross almost anything the Antarctic could throw at them, and would bounce along through soft snow and challenging sastrugi fields.

The tents belonged to a Taiwanese team who were being supported by members of an Icelandic company, Arctic Trucks. The group had been dropped off around two degrees away from the Pole and were steadily making their way back to the station. The Taiwanese guys welcomed me like a long-lost friend while snapping away with their cameras.

'Hi, Captain Rudd, Captain Rudd, welcome, welcome,' one of the Taiwanese guys said, hugging me and patting me on the back.

They seemed to know who I was and what I was doing, and one guy even said that we were friends on Facebook. 'Come in, come in, have some coffee, chocolate, come in, get warm,' they kept saying.

They were extremely friendly, but the language barrier meant that I was struggling to explain that I couldn't take any food or drink because of the rules of my expedition. As we chatted away, I could feel my body temperature plunging and told my new-found friends that I had to get a move on, although I'm not entirely sure the message got through. I smiled graciously and tried not to be rude, continuing to wave at them until I was virtually out of sight.

Clear of the Taiwanese camp, I pressed on, making good progress. I had a good 20 mph tailwind pushing me along, and the snow was hard and relatively flat. Although my face mask was doing its job in protecting my ulcerated mouth, I was still slipping into periods of intense pain. But my morale remained high, mainly because I could tell that I was now eating up the miles. Over the next few days the good weather remained, and my mileage was increasing all the time, even though I had still not reached the summit of the area I was in.

Now I began to rack up a series of mileage records, and a week on from leaving the South Pole I skied just over 23 nautical miles in a single day. My pulk was lighter, my skins had more glide, and the tailwind was proving a godsend, on occasions gusting up to 50 mph. I was effectively being blown along. I was still finishing each day exhausted, but the expedition had become really enjoyable. My mouth was getting a little better and, although I missed my family, I relished my solitude. I still had a long way to go but I knew that, barring any catastrophes, I should complete the crossing. The pressure of failure had eased and it was truly liberating. It also began to dawn on me that I would miss this place once I was back home in the UK, and so I vowed to savour every last minute of it.

On the tenth day out from the Pole I got my first glimpse of the snowy peaks of the Transantarctic Mountains. The range was a good three days away still, but it was great to have a real feature to aim for, rather than just a flat line on the horizon. In good weather I didn't even need to look at my compass. I could just pick one of the peaks and ski on that bearing. I marvelled at how different the final leg of the expedition was compared to the first. Skiing away from the Pole was fun, while the first half of the expedition to the Pole had almost broken me. Hour by hour, as I skied forward, the beauty of the dramatic mountain range was gradually unveiled. It was an exhilarating and emotional time.

A few years earlier I had reached a similar point with Henry, but coming from the other direction, and back then I remember a lump in my throat when I first encountered the mountains. The same thing had happened on SPEAR, and I was almost as emotional this third time. This mountain range offered salvation; when I reached it, I basically had a glacial ski run down to the Ross Ice Shelf and my pick-up point. But best of all, I had crossed 87 degrees south for the second time on this trip, but this time heading north. It was definitely a bit odd to be counting the degrees in reverse, and I had to remind myself sometimes that – for example – S87° 40' meant I had 40 nautical miles to go to reach the next degree, not 20.

Although I still had 100 miles to go until the finish, I felt the time had come for me to raid my five-day reserve rations. My body was now craving food all the time and I never felt satiated, even after a large meal. At night when I closed my eyes and my mind wandered just before I fell asleep, I would think about steak and chips, curry, roast potatoes, and good hearty home-cooked meals. It was almost like a form of self-torture. I would sometimes salivate at the thought of real food or even a decent cup of coffee made with real milk. Knowing that it was beyond my reach made it all the more desirable.

Christmas Eve arrived, almost unexpectedly, on Day 53, and I decided to change my routine slightly so that – come Christmas Day – I would be able to call home and speak to my family during their daytime. With that in mind I decided to try and get to bed early, catch a few hours' sleep, wake at 1 a.m., have my breakfast and get on the move by 2.30 a.m. The plan went like clockwork and I managed to ski for thirteen hours until 3.30 p.m. and complete a satisfying 20.4 nautical miles. I rested for the next five hours and set off again at 9 p.m. on Christmas Eve, skiing all through the night until 5 a.m. on Christmas Day, when I stopped, put my tent up, got some sleep, and then called home just as my family back in Hereford were about to sit down for Christmas dinner.

Before I left for Antarctica, I had written a letter to the family, placed it in a sealed envelope and asked Lucy to read it out on Christmas Day. Lucy had opened the letter and begun reading it, but got a bit too emotional, so it was passed to my eldest daughter, Amy, who also broke down in tears before finally Sophie, who is the most emotionally robust member of the family, took over and began to read it out. In the letter I told Lucy and my three children just how proud I was of each of them. I then cited various moments in their lives that had been huge milestones for them and how I was the proudest father in the world.

Just before I called, I imagined them all sitting around the dinner table, ready to pull crackers and enjoy a lovely roast turkey, and I was almost overcome by guilt. Once again I was away from my family for Christmas, the third time in six years, on top of all the many I had missed before that during my time in the SAS.

When we spoke, I could clearly hear the emotion in their voices. It was tough, very tough for me. It was nice to connect with the family, especially on Christmas Day, but it was also an emotional drain.

I was up early on Boxing Day and managed to reach the top of the Leverett Glacier, which is like a huge mountain pass, up to three miles wide in some places, covered in deep, soft snow and flanked by vast mountainous peaks. It was quite steep in places and I could ski down in snowplough style, throwing in a few turns, but more often than not I would end up in a bit of a heap as the pulk began to gather speed and crash into the back of my legs. The further I progressed, the worse I seemed to fare, and big wipeouts were becoming increasingly common. Although it was something of a relief to be sliding downhill, I had to be careful not to do something daft like twist a knee or dislocate a shoulder.

I ended the first day on the Leverett by skiing just over 24 nautical miles and descending down to 5,000 feet. That was also the day that I was told Colin had finished. It later transpired that he had completed the final 72 miles in an epic thirty-two-hour push. It was an impressive feat of physical and mental endurance, and I wasn't surprised that he had gone to such extreme lengths to get to the finish line first. I had chosen not to race right from the beginning, and I had no regrets about my decision.

By the end of Day 55 I had skied just over 29 nautical miles in fourteen hours, a personal expedition mileage record. As I prepared my food that night in the tent, I reflected on what I had achieved and actually felt quite satisfied with my performance. Ultimately, finishing just two days behind a professional endurance athlete, sixteen years my junior, who undertook these huge expeditions for a living, wasn't so bad.

On Day 56 I skied off the Leverett Glacier and onto the Ross Ice Shelf. It was like entering another world; a vast, white, unending wilderness for as far as the eye could see. It was truly amazing to be standing on it once more. From a distance, the ice shelf looks perfectly flat, but the reality is very different; there are lumps, bumps, ice ridges and deep crevasses criss-crossing the entire area. I skied on for about five or six miles, heading to the waypoint that ALE had provided, which is just an arbitrary point out on the shelf, clear of the mouth of the glacier.

As the mountains began to recede behind me, I spotted Colin's red tent and a cache of green 45-gallon aviation fuel drums in the far distance. I picked up my pace and, within an hour or so, had arrived at the campsite, at around 4.20 p.m. Having dropped off my pulk and unclipped my skis, I walked over to Colin's tent, tapped on the side and stuck my head in to say hello. Colin wasn't there, so I pulled my head out and had a look around, but he was nowhere to be seen. His pulk was by his tent, as was all of his gear apart from his skis and poles. I scanned the horizon back towards the Leverett Glacier and out over the Ross Ice Shelf, but – even though visibility was good and I could see for several miles in every direction – there was no sign of him.

Rather than stand around getting cold, I set up my tent some 20 yards away and then called in to ALE to let them know that I had arrived so they could begin planning the pick-up. As I relayed the information, I asked if they knew where Colin was, but they were unaware of his movements. It was a concern for them, because that meant Colin was effectively missing; he hadn't told them he was planning to go anywhere and his tracking beacon wasn't activated.

Colin eventually appeared later that evening.

'Where have you been, mate?' I said, shaking him by the hand after congratulating him on his successful expedition.

'Oh, I've just been out taking a few photos.'

'OK, cool,' I responded. 'You went a long way to take a few pictures.'

Colin shrugged and clearly didn't want to talk about his excursion. I thought it was odd, but couldn't work out what else he might have been doing. I was really impressed with what he had achieved and I had to take my hat off to him. He looked physically exhausted, his face was covered in tape to protect his cold injuries, and I could see that he – like me – had lost a lot of weight.

He then hugged me: 'Fantastic job, man,' Colin said, sounding like a true American. 'I'm so pleased you are here – me and you together, the first two people on the planet to ski across the Antarctic landmass in this style.

'It was so tough, Lou,' he went on. 'Far harder than I could have ever imagined. There were so many times I thought about giving up. Knowing that you were also on the ice and coping with those conditions helped me enormously, so I have to thank you for just being out there.'

And he was right: despite the fact that I was trying to blank out that he was also out there, I think subconsciously we drove each other on. Knowing there was someone else doing the same thing and coping with everything that was being thrown at them showed that it was possible.

And I suppose it was at that point, standing on the frozen ice shelf, just the two of us, reminiscing on all we had been through, that the enormity of the achievement finally began to sink in. We were the first two people in history to ski across the Antarctic landmass solo and unsupported, using muscle power alone. No kites, no wind sails, no resupplies of food or equipment. This hadn't been a return trip where you would have the luxury of laying depots for yourself on the outbound leg, thus dramatically lightening your load every few days, and then doing the return leg dashing from depot to depot with a half-empty pulk. This was a straight A to B expedition using pure muscle power and being completely self-sufficient for two months in the most hostile environment on the planet. There is no purer form of polar travel. Two unbroken ski tracks of over 900 miles lay behind us, stretching all the way back to the far side of Antarctica.

I say it began to sink in, but I'm not sure it really did, not for a few days anyway. The biggest emotion that I felt at the time was relief. Relief that I had completed the crossing and not failed. Relief that I had got through it without too many injuries, and relief that after fifty-six days of skiing, without a single rest day, and hauling and battling my way through almost impenetrable fields of sastrugi and deep, soft snow, it was finally over and I would now be able to relax – for a few days, anyway.

As I stood there taking it all in, Colin then turned to me and said: 'How much food have you got left? I'm pretty much out.'

'There should be a cache of food buried near the oil drums,' I said.

'Yeah, found that, and I've already started eating some of it, but I've saved a bit for you.'

I laughed loudly, 'Don't worry, mate. I've got around five days' supply left, so we can share that – split it fifty-fifty.'

A cool-box of food had been deposited by ALE the season before, in preparation for Ben Saunders's solo attempt to cross the continent, in case he was low on supplies when he arrived at the finish. Despite being buried in the ice for a year, the contents – consisting of cheeses, chocolate, salami, sweets and coffee – were a welcome addition to my few remaining freeze-dried meals. Ben's misfortune in not completing his trip had been our good luck.

Later that evening, ALE called us back to explain that bad

weather meant the pick-up would be delayed by at least three days. I filled the hours with phone calls to family, friends, supporters and sponsors. I watched a couple of movies, slept and chatted to Colin, who was downbeat about the delay and urged me to speak with ALE to try and persuade them to come earlier. I think he thought that, due to my prior expeditions, I would have more sway with ALE than he did. But I had to explain several times that the guys at UG were not going to risk an aircraft and two pilots just to get us back a day or so earlier. On a similar journey in 2013, in thick cloud, a Twin Otter crashed into the side of Mount Elizabeth in Antarctica, and all three crew members were killed. A stark reminder of the perils of operating aircraft on this unforgiving continent. We would both be home soon enough and, for me, it was important to drink in the very last images of one of the most out-of-this-world landscapes on the planet.

Colin did later go on to claim that he had waited for me at the finish in an act of sporting chivalry, but I'm not sure he had a choice. With us finishing so close to one another, ALE were never going to conduct two high-risk twelve-hour round trips within a couple of days to collect us separately.

The pick-up eventually came on 30 December at around 1700 hours. Colin and I were sitting on our pulks, all packed and ready to go, playing 'who can spot the aircraft first'. The Twin Otter seemed to circle around for ages, obviously trying to find a relatively flat bit of ice, before eventually landing. The skill of these pilots always astounds me. The crew jumped out and greeted us with hugs and a bottle of champagne. The plane refuelled, the kit was stowed in the aircraft's hold, and within forty minutes we were airborne, with Colin and me squished in the back, along with the empty fuel drums. The team at Union Glacier had prepared some packed lunches of smoked salmon wraps, biscuits and crisps, washed down with the champagne, which has to rate as one of the finest lunches I've ever had.

Going without for so long can give you a whole new perspective on even the simplest of things.

The cabin inside the Twin Otter was very noisy and so, after a short while, Colin and I gave up on trying to have a conversation and instead concentrated our efforts on finishing the champagne. The booze quickly rushed to our heads and, as I looked out of the window, down on the vast landscape effortlessly slipping by far below, I was already feeling a sense of loss and dislocation. I had spent so much time fully immersed in that environment and now I felt I was suddenly being torn away from it. I had an overwhelming urge to get out of the plane and be back on the ice, just for a little longer.

The aircraft stopped off at the airstrip at Thiel Mountains to refuel, and we eventually landed back at Union Glacier around 11 p.m. that evening. What had taken me the best part of two months to ski, we had just flown over in around five hours.

Virtually the entire ALE staff – along with other expeditioners – came out and surrounded the plane to welcome us back. The champagne flowed once again and we were both treated like conquering heroes. A hot meal was waiting for us and everyone was keen to hear about the highs and lows of the expedition. I carried on eating and drinking until around 5 a.m. when, after a short bucket shower, I finally crashed onto a cot-bed inside one of the ALE staff tents. I called home the following morning and spoke to Lucy. It was great to hear her voice again and, as we chatted, she said that there seemed to be lots of media interest in the expedition. I thought there might be some, but I was completely taken aback by the level of coverage. From the moment I ended my call to Lucy, the phone barely stopped ringing. Wendy was inundated with requests and did her best to schedule the interviews. There was lots of British interest, but I was also receiving calls and emails from news organizations around the world who wanted to know every detail of what had taken place.

Our three-day stay at UG included New Year's Eve, which

was celebrated in style with lots of drinks and an outside BBQ at -20°C. It was a fantastic end to the expedition, but all a bit surreal.

On 3 January 2019 I flew back into Punta Arenas and civilization proper. It was raining and very drab, and the glorious blue, uncluttered skies of Antarctica seemed a long way away. Over the next couple of days, I began packing away my equipment and preparing it for airfreight when Wendy got a call from the US saying they wanted to fly me up to New York on an all-expenses-paid three-day trip to do a round of TV interviews. I was amazed at how the expedition had captured the public's imagination in such a powerful way, and of course I agreed. It was exactly the sort of coverage the sponsors were after, and I also wanted the opportunity to tell my side of the story.

I had a busy three days of media calls with CBS and HBO. On one show I was introduced to Malala Yousafzai, the young Pakistani girl who had been shot in the head by Taliban gunmen and who went on to win a Nobel Peace Prize. The trip also included a visit to the New York headquarters of the prestigious Explorers Club and, during a personal tour with the club president, I was invited to become a member. It was a true privilege to become part of such a historic organization, which counts all the early polar pioneers, US presidents and many Apollo astronauts among its membership.

It was all great fun being treated like royalty, but the longer it went on, the more surreal the whole experience became. Just a few days earlier I had been on my own, completely self-sufficient, hauling a pulk across a snowbound wilderness. I had gone from one of the most remote places on the planet to one of the busiest, and it took some getting used to. I later discovered that the huge media interest had been generated by the *New York Times*, which had spun the two expeditions as a race, and while Colin and I had been away, the story had been building and building. In the end, something like 200 articles worldwide had been written about the expeditions, with estimated coverage views of 10 million.

Despite the fun and glamour of being a media star – albeit briefly – it was time to get home, and I eventually flew into Gatwick on 9 January, where I was greeted by friends, family, sponsors and the media. I had ten minutes of hugs and kisses with my family before being whisked off to face the press once more. It was all a bit strange, to put it mildly. I had spent all of my SAS career avoiding the media, and now I was expected to give interviews to anyone who wanted them. I spent five hours at the airport going live on Sky News, BBC and ITV.

I arrived home later that evening, marvelling at the greenery of the Herefordshire countryside.

The comfort of my own bed was immeasurable, but I couldn't dismiss the feeling of being slightly lost. I had developed a spiritual connection with Antarctica and I was missing it already. Of all the emotions I thought I would experience when I returned home, this was the most unexpected.

After all our time apart, sharing this with my wife clearly wasn't an option, so on my month off before I returned to work, I focused on the impending media aftermath, which in itself was at times somewhat overwhelming. I also spent time producing a lengthy post-expedition report for the Army. Already my monastic daily routine on the ice began to feel like a distant memory.

When I did return to work, I spent the next five months touring the UK, visiting over a hundred schools and cadet centres, talking about Army Adventure Training and my Antarctic expeditions.

Many miles were covered each week; it was thrilling to have been given the opportunity to engage with over 24,000 young people. Hopefully some at least will now know a little more about Antarctica and Britain's proud polar history.

In May 2019 I was honoured to be made 'Explorer of the Year' by the Scientific Exploration Society, and I also met members of

Winston Churchill's family, who had heard about me listening to audiobooks about their great-grandfather as I skied across the continent.

I missed the simplicity of my solitary fifty-six days on the ice. Yes, I'd completed my goal, but it had always been about more than that for me. It was about the love of being on the journey; about the privilege and experience of being alone in Antarctica and marvelling at its scale and its raw, natural beauty; about trying to survive and travel safely within an environment that feels like it could take your life in an instant, a place where it seems as if humans, indeed any form of life, are just not meant to be.

It is often said that once you've been to Antarctica all you do is dream about returning – but there is often a heavy price to pay. Robert Falcon Scott, Ernest Shackleton, and my mentor and great friend Henry Worsley all returned to the great southern continent while endeavouring to explore it further. But their desire to push the boundaries of what was achievable ultimately cost them their lives.

You don't just visit Antarctica, it holds you fast in its grip. I know that it's only a matter of time before I once again answer the siren call of the Great White Queen. After all, Endurance is in my blood. Onwards.

ACKNOWLEDGEMENTS

A successful expedition to Antarctica is dependent on a small army of people coming together to help with finance, equipment, logistics expertise, advice and a million other things. I owe so much to so many people that it would require another whole chapter to list their names. Just about everyone who knows me has helped in some small way by sharing my excitement and enthusiasm, cajoling me when I occasionally questioned the sanity of my adventures, or kindly listening for hours on end to my thoughts. But, as with any venture, there are always those whose input is crucial, and without their efforts none of my expeditions would have seen the light of day.

First and foremost, I would like to thank my wife Lucy and my three wonderful children, Amy, Sophie and Luke, for your patience, understanding and a few missed Christmases. Knowing you were home waiting for my return always gave me the greatest motivation to keep pushing one ski in front of the other. I am so proud of all of you.

To Wendy Searle, who has been absolutely critical over the last few years with her support and unstinting belief in all of my crazy plans, as both my expedition manager and adventure companion. Your enthusiasm and drive have inspired me to keep going, even during the darkest times. I've lost count of the number of times I've called you from Antarctica in the dead of night to seek your wise counsel. It's been amazing to watch you transition from ardent supporter to expeditioner yourself and achieve your own incredible goals. You just make things happen.

To Henry Worsley, for believing in me and giving me that first opportunity to be part of an Antarctic expedition, which has led to everything that followed. You inspired me; through your own actions you showed me how to be a good leader, and I know you are still looking out for me now. No man could ask for a greater travelling companion. And to his wife Joanna who, despite her own challenges, has continued to support me and provide her blessing for this book, which means so much.

To the SPEAR team, Ollie, Alex, Chris, Al and James. No man could ask for greater teammates. We have a friendship for life that has been forged through overcoming shared adversity, and celebrating triumph. To all of the team at Antarctic Logistics and Expeditions, a huge thank you for all of your professional and friendly support over the years. Particularly Steve Jones, for all of your planning support and advice; legendary polar guide Hannah McKeand for keeping me grounded and humble and showing me alternative ways of doing things; and Martin Rhodes for the red wine, laughs and medical support. Also from the polar world, Lars Ebbesen: without your experience and guidance, Greenland would never have happened. And thanks again for the heads up on the piteraq that almost ended our trip!

There have been so many wonderful sponsors over the years and the sheer elation I have experienced when each and everyone one of you has agreed to support in whatever way you can will live with me forever. To Martin Brooks and Ian Holdcroft from Shackleton, thank you for being the most wonderful expedition partners. Your support for Spirit of Endurance was instrumental in its success and to be a small part of your visionary brand is a privilege. To the Ulysses Trust, Team Army, SAS Regimental Association, The Clocktower Foundation, SSVC, Army Sports Lottery, the Reserve Forces' and Cadets' Associations, Compass Underwriting, Robin Bacon and Richard Hackett at ABF The Soldiers' Charity, Kevin Mann at Level Peaks Associates, 7R Performance, Bourne Hall, Gus Majed, John Jones, Pete and George Masters at Skinzophrenic Tattoos, Mary at Expedition Foods,

Alex at Acapulka, Richard Crook at Airbus, the Scientific Exploration Society, the Explorers Club, Professor Chris Imray and many, many more. Thank you all for everything you have done in making all of this a reality. I am eternally grateful.

To the amazing support team behind the publication of this book: Sean Rayment for writing support and planting the seed to do the book in the first place; Julian Alexander, my literary agent; and Ingrid Connell, my publisher at Pan Macmillan. It's been a long journey, but we got there in the end.

From the military, I'd like to thank my former commanding officers Neil Grant and Shaun Chandler for your unstinting support for SPEAR and Spirit of Endurance respectively. Allowing me the time and space to pull these mammoth projects together was critical to their success, and believing in my ability to safely deliver them meant so much. And to the next level up, the brigadiers and generals who ultimately had to sign off on these high-risk endeavours, thank you for believing in me. Particularly to General Tyrone Urch for your most generous grant support and saving the day with the Brexit debacle, and to Major General Paul Nanson for signing on the dotted line for my most recent solo endeavour. To the team at Army Adventure Training Group, thank you for the planning, equipment and grant support on all three Antarctic expeditions, particularly Nick Richardson, Graham Cook, Cliff Pearn and Chris Coates. All of you represent the absolute best of the British Army.

Finally, a thank you to three of the most incredible military units in the world, which I have had the privilege to serve in: the Royal Marines, the Special Air Service and the Parachute Regiment. Your values, standards and ethos have made me the man I am.